SYMMETRY IN MOLECULES AND CRYSTALS

First published in 1989 by
ELLIS HORWOOD LIMITED
Market Cross House, Cooper Street,
Chichester, West Sussex, PO19 1EB, England
The publisher's colophon is reproduced from James Gillison's drawing of the ancient Market Cross, Chichester.

Distributors:

Australia and New Zealand:
JACARANDA WILEY LIMITED
GPO Box 859, Brisbane, Queensland 4001, Australia

Canada:
JOHN WILEY & SONS CANADA LIMITED
22 Worcester Road, Rexdale, Ontario, Canada

Europe and Africa:
JOHN WILEY & SONS LIMITED
Baffins Lane, Chichester, West Sussex, England

North and South America and the rest of the world:
Halsted Press: a division of
JOHN WILEY & SONS
605 Third Avenue, New York, NY 10158, USA

South-East Asia
JOHN WILEY & SONS (SEA) PTE LIMITED
37 Jalan Pemimpin # 05–04
Block B, Union Industrial Building, Singapore 2057

Indian Subcontinent
WILEY EASTERN LIMITED
4835/24 Ansari Road
Daryaganj, New Delhi 110002, India

© **1989 M.F.C. Ladd/Ellis Horwood Limited**

British Library Cataloguing in Publication Data
Ladd, M. F. C. (Marcus Frederick Charles)
Symmetry in molecules and crystals
1. Crystals. Symmetry
I. Title
541.2′2

Library of Congress Card No. 88–38190

ISBN 0–85312–255–5 (Ellis Horwood Limited)
ISBN 0–470–21376–0 (Halsted Press)

Printed in Great Britain by Hartnolls, Bodmin

SYMMETRY IN MOLECULES AND CRYSTALS

M. F. C. LADD, B.Sc., M.Sc., Ph.D., D.Sc.
Reader in Chemical Crystallography
University of Surrey

ELLIS HORWOOD LIMITED
Publishers · Chichester

Halsted Press: a division of
JOHN WILEY & SONS
New York · Chichester · Brisbane · Toronto

Corrigenda:-

SYMMETRY IN MOLECULES AND
CRYSTALS
(M. F. C. Ladd)

Page 69	*Table 3, in the row beginning* 'Inverse diad', *the third column should read* ' $0 \perp$ projection $\quad 0 \parallel$ projection'.
Page 158	*Line 3 after Fig. 5.14,* after *symbol insert* $\rightarrow$
Page 176	*Scheme at top:* *Delete overbars on uppermost and lowermost 4s in the central column.*
Page 185	*Scheme near bottom:* *Central part of top line to read:* $\ldots \quad \overline{y}, x - y, z \quad \ldots$
Page 202	*Eq (6.17), on the right-hand side,* ' $\dfrac{h+k}{2}\Big]$ ' *should read* ' $\dfrac{h+k}{2}\Big)\Big]$ '.
Page 203	*Eq (6.20), on the right-hand side,* ' $l/2)]$ ' *should read* ' $l/2))]$ '.
Page 208	*Problem 10, on the right-hand side,* ' $lz_m)$ ' *should read* ' $lz_m)]$ '.
Page 264	*Solution 9,* ' $f_{j,\theta} \sin 2\pi(hx_j \ldots$ ' *should read* ' $f_{j,\theta}[\sin 2\pi(hx_j \ldots$ '.

The Author apologizes for these errors.

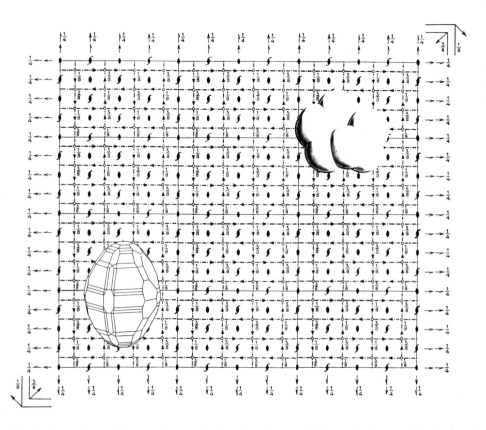

Frontispiece — Space-filling model of a molecule of S_8 and a crystal of S_8 on a background of a drawing of space group <u>Fddd</u>, that of S_8.

Table of contents

To Anthony and Nicholas

whose father
I have the privilege to be

Preface

The study of crystal and molecular symmetry is often considered difficult by those meeting it for the first time. There are, I believe, two reasons for this apparent problem. First, crystals and molecules, unlike the more familiar plane figures, are three-dimensional bodies, and it is not possible to take in simultaneously all parts of a body and see them in relation to the whole. In two dimensions, with a plane hexagon, for example, it can be done easily, and we accept the symmetry of a hexagon without difficulty. The second reason, not unrelated to the first, lies in the unfamiliarity of three-dimensional relationships rather than in their intrinsic difficulty, despite the fact that we live in a world of solids and symmetry.

The purpose of this book is to try to present crystal and molecular symmetry in a straightforward, practical manner to all those who come newly to this subject. It has evolved from courses given over many years, at graduate and postgraduate levels, to crystallographers, chemists, chemical physicists, metallurgists and physicists, and makes liberal use of stereoviews and other aids to understanding symmetry. Each chapter concludes with a set of problems, for which detailed solutions are provided. It is informative to study everyday objects in relation to symmetry: a glass or a beer mug; a chair or a table; a Dalmatian dog or a Dobermann; a tiled floor or a brick wall. In this way, symmetry becomes clearer—it is not only about crystals and molecules!

There are two symmetry notations in current use, one due to Hermann and Mauguin and the other due to Schoenflies. The majority of the book emphasizes the Hermann–Mauguin notation, because it is the more generally useful, but some aspects of molecular symmetry, and particularly the sections on group theory, are presented in the Schoenflies notation. This procedure is traditional, although it is to be hoped that eventually a single symmetry notation will be adopted. Some suggestions in this direction are given in two or three places in the book.

The geometry of crystals may be studied beginning either with lattices or with the description of the external symmetry of crystals. The former approach has advantages in a short lecture course, and in getting to grips readily with X-ray diffraction from crystals. However, the traditional morphological approach has been chosen here because, in that way, the book should have a more general appeal, and its pace is slower. Time is needed in order to assimilate the concepts of symmetry and to consolidate them into a practical, working knowledge of the subject. It is hoped that the book will be useful to all those who are meeting symmetry for the first time, whatever their specialization.

I am most grateful to Don Rogers, Emeritus Professor of Chemical Crystallography at Imperial College of Science and Technology, London, for reading the book in manuscript, and for making several valuable comments and suggestions, which have been adopted in the book. I am pleased to acknowledge the courtesy of the publishers listed hereunder for their permission to reproduce the illustrations that have been attached to their names. Finally, I am indebted to Ellis Horwood MBE and the staff of the firm that bears his name for their assistance to me in bringing this work to a state of completion.

Plenum Press, *Structure Determination by X-Ray Crystallography*, M. F. C. Ladd and R. A. Palmer (2nd edn, 1985): Figs 2.1–2.11, 2.15, 2.21, 2.23, 3.4, 3.6–3.10, 3.12, 3.22, P3.1, 4.28, 5.2, 5.4, 5.6, 5.7, 5.9, 5.11, 5.12 5.14–5.17, 5.19, 5.21, 5.25, 6.6–6.10, A1.1, S2.2, S5.4, S5.7.

International Union of Crystallography, *International Tables for X-Ray Crystallography*, Volume I (1965): Figs 5.22–5.24, 5.27, 5.28(b), 5.29–5.32.

Macmillan and Co., *Crystallography and Practical Crystal Management*, A. E. H. Tutton (1911): Fig. P2.2, crystal drawing in Figs 3.11a–h, 3.13a,b, 3.17, 3.18c,d, 3.19, 3.20a,b,d, 3.21.

McGraw-Hill Book Co., *Introduction to Crystal Geometry*, M. J. Buerger (1971): Figs 3.11i,j, 3.13c, 3.18a,b, 3.20c.

Longman Group Ltd, *An Introduction to Crystallography*, F. C. Phillips (4th edn, 1971): Figs 3.25, 3.26.

National Physical Laboratory, *Internal Report*, R. J. Bell and P. Dean (1969): Figs 6.1, 6.3 (Crown Copyright).

A. Oosthoek's Uitgeversmaatschappij NV, *Symmetry Aspects of M. C. Escher's Periodic Drawings*, C. H. Macgillavry (1965): Fig. 5.3.

G. Bell and Son Ltd, *The Crystalline State*, Volume I, Sir Lawrence Bragg (1949): Fig. 6.4.

Interscience Publishers, *Advanced Inorganic Chemistry*, F. A. Cotton and G. Wilkinson (2nd edn, 1966): Fig. 7.2.

Farnham 1988

M. F. C. Ladd

List of symbols

The symbols here, mostly standard, are used generally throughout the book. Other notation has been described in its localized context.

A'_{hkl}, B'_{hkl}	Components of the structure factor, measured along the real and imaginary axes respectively
A	A-face-centred unit cell; absorption correction factor
Å	Ångstrom unit ($1\ \text{Å} = 10^{-10}\ \text{m} = 10^{-8}\ \text{cm} = 10\ \text{nm}$)
a, b, c	Unit-cell vectors parallel to the x, y and z axes respectively
a, b, c	Unit-cell dimensions parallel to the x, y and z axes respectively; intercepts made by the parametral plane on the x, y and z axes respectively; glide planes with translational components $a/2$, $b/2$ and $c/2$ respectively
a*, b*, c*	Reciprocal unit-cell vectors parallel to the x^*, y^* and z^* axes respectively
a^*, b^*, c^*	Reciprocal unit-cell dimensions parallel to the x^*, y^* and z^* axes respectively
B	B-face-centred unit cell
C	C-face-centred unit cell
C_R	Rotation axis (Schoenflies notation)
$\cancel{C}$	'not constrained by symmetry to equal'
c	Speed of light *in vacuo* ($2.9979 \times 10^8\ \text{m s}^{-1}$); see also a, b, c above
D_R	R 2-fold rotation axes perpendicular to C_R (Schoenflies notation)
D_m	Experimentally measured crystal density
d	Differential operator
d	Interplanar spacing
d_{hkl}	Interplanar spacing of the (hkl) family of planes
d*	Reciprocal space vector
d^*	Distance in reciprocal space

$\mathbf{d}^*_{hkl}$	Vector from the origin to the hkl reciprocal lattice point
d^*_{hkl}	Distance from the origin to the hkl reciprocal lattice point
$\mathbf{e}$, exp	Exponential factor (2.71828 ...)
e	Electron charge (1.6022×10^{-19} C)
F	All-face-centred unit cell
F_{hkl}	Structure factor for the (hkl) family of planes referred to one unit cell
$f_{j,\theta}$	Atomic scattering factor for the j atom in the unit cell
H	Hexagonal (triply primitive) unit cell
(hkl), $(hkil)$	Miller and Miller–Bravais indices respectively, associated with the x, y and z or x, y, u and z axes (any single index except the last containing two digits has a comma placed after it)
$\{hkl\}$	Form of (hkl) planes
hkl	Reciprocal lattice point corresponding to the (hkl) family of planes
h	Planck constant (6.6256×10^{-34} J s)
I	Body-centred unit cell; intensity of transmitted X-radiation
I_0	Intensity of incident of X-radiation
I_{hkl}	Intensity of X-radiation from the (hkl) family of planes, referred to one unit cell
$\mathbf{i}$	$\sqrt{-1}$; an operator that rotates a vector in the complex plane through $90°$ in an anticlockwise manner; the imaginary
K	Reciprocal lattice scale factor; characteristic (K) X-radiation
L	Lorentz correction factor
M_r	Relative molecular mass
m	Mirror reflexion plane (Hermann–Mauguin notation)
n	Glide plane, with translational component $(a+b)/2$, $(b+c)/2$ or $(c+a)/2$
$\not{p}$	Primitive unit cell (one dimension)
p	Primitive unit cell (two dimensions); polarization correction factor
P	Primitive unit cell (three dimensions)
$\mathbf{R}$	Rotation matrix
R	Rhombohedral unit cell; rotation axis of degree R (Hermann–Mauguin notation)
$\bar{R}$	Roto-inversion (inversion) axis of degree R (Hermann–Mauguin notation)
S_R	Alternating symmetry axis of degree R (Schoenflies notation)
$\mathbf{t}$	Translation vector
u	Atomic mass unit (1.6606×10^{-27} kg)
$[UVW]$	Zone axis; direction from the origin to lattice point U, V, W
$\langle UVW \rangle$	Form of zone axes or directions
$\mathbf{x}$	Triplet x, y, z written as a vector
x, y, z	Reference axes; fractional coordinates
x_j, y_j, z_j	Fractional coordinates of the j atom in a unit cell
X, Y, Z	Coordinates in absolute measure (nm or Å)
$[x, \beta, \gamma]$	Line parallel to the x axis and intersecting the y and z axes at β and γ respectively
(x, y, γ)	Plane normal to the z axis and intersecting it at γ
Z	Number of formula-entities in a unit cell

α, β, γ	Angles between the pairs of unit-cell edges **bc**, **ca** and **ab**; angles of general rotation
$\alpha^*, \beta^*, \gamma^*$	Angles between the pairs of reciprocal unit-cell edges **b*c***, **c*a*** and **a*b***
Γ	Irreducible representation
θ	Bragg angle
θ_{hkl}	Bragg angle for the (hkl) family of planes
λ	Wavelength
μ	Linear absorption coefficient
ρ	Radius of stereographic projection
σ	Mirror reflexion plane (Schoenflies notation)
ϕ	Internormal angle; atomic wave function
χ, ψ, ω	($\cos \chi$, $\cos \psi$, $\cos \omega$) direction cosines of a line with respect to the x, y and z axes, respectively

1

Symmetry Everywhere

"Beauty is truth, truth beauty,"—that is all
Ye know on earth, and all ye need to know

John Keats, 1820: *Ode on a Grecian Urn*

1.1 LOOKING FOR SYMMETRY

Few of us have difficulty in recognizing symmetry in two-dimensional objects such as the outline of a shield, a Maltese cross, the three-legged emblem of the Isle of Man, a five-petalled Tudor Rose or the six-pointed Star of David. But it is a rather different matter when, as with much of this book, we are dealing with three-dimensional objects. The difficulty stems partly from the fact that we can see all parts of a two-dimensional object simultaneously, and thus see the relation of the parts to the whole; but we cannot do that for a three-dimensional object. Secondly, while some three-dimensional objects, such as flowers, pencils and architectural columns, are simple enough for us to visualize and to rotate in our mind's eye, few of us have a natural gift for mentally perceiving and manipulating more complex three-dimensional objects. Nevertheless, the art of doing so can be developed with suitable aids and practice, and especial care has been taken in this book to help the reader to gain fluency and confidence in handling three-dimensional objects. If, initially, you have such problems, take heart. You are not alone, and, like many before you, you will be surprised at how swiftly the required facility can be acquired. Engineers, architects and sculptors may be blessed with a native three-dimensional visualization aptitude, but they have learned to develop it—particularly by making and handling models.

Standard practice in the past was to reduce three-dimensional objects to one or more two-dimensional diagrams (projections or elevations): it was cheap, well suited to reproduction in books and less cumbersome than handling three-dimensional models. In this book we shall continue to use such two-dimensional representations, but to rely on them exclusively only tends to delay the acquisition of a three-dimensional visualization facility. Fortunately, we can now also use stereoscopic image pairs. Such illustrations are a great help but, because they provide a view from only one standpoint, they are not always quite the equal of a model that can be manipulated by hand. Some notes on stereoviewing are given in Appendix 1.

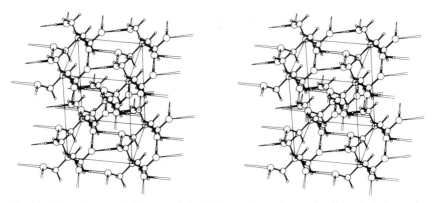

Fig. 1.1—Crystal structure of pentaerythritol in the environs of one unit cell. Circles in increasing order of size, represent H, C and O atoms; hydrogen bonds are shown by double lines.

As an example, Fig. 1.1 shows part of the crystal structure of pentaerythritol, $C(CH_2OH)_4$, which consists of discrete molecules. The volume depicted is that of the unit cell (the three-dimensional repeating unit) and includes the bordering molecules. We can cover either half of the figure and study the structure as a two-dimensional representation. Although the figure has been drawn carefully, with depth indicated by tapered bonds, it does not convey nearly as much information as that which utilizes the stereoscopic effect obtained with the pair of drawings: a good stereoview, seen correctly (Appendix 1), conveys a convincing three-dimensional impression.

Four quite different objects are illustrated in Fig. 1.2. At first, there may not seem to be any connexion between a Dobermann bitch, a Grecian urn, a molecule of 3-chlorofluorobenzene and a crystal of potassium tetrathionate. Yet each is an example of reflexion symmetry: a symmetry plane, symbol *m* (mirror), can be imagined to divide each object into two halves that are related as is an object to its mirror image. If we could perform the operation of reflecting the two halves across the mirror plane, then the object would appear unchanged. If we view the Dobermann from the side its mirror symmetry is not evident, though it is still there, of course. If, however, we stand the side view in front of a mirror, then the Dobermann and its image (Fig. 1.3) now show together the same type of symmetry, but with the plane of symmetry outside the animal.

We can pursue our analogies a little further. The Dobermann, beautiful animal that she is, if scrutinized carefully will be seen not to have perfect *m* symmetry. Again, only the outline of the urn conforms exactly to *m* symmetry; the decorations do not show that symmetry. Under a microscope, real crystals can be seen to have minute flaws that are not in accord with the symmetry of the conceptually perfect crystals shown by drawings, such as that of Fig. 1.2(d). If we consider the internal symmetry of the Dobermann it will certainly not be *m*. Crystals behave in a similar manner: alum, $KAl(SO_4)_2 . 12H_2O$, crystallizes as octahedra, but its internal symmetry is less than that of an octahedron.

If we seek symmetry examples around us we soon encounter repeating patterns, as well as one-off (finite) bodies. For example, consider the tiled floor of Fig. 1.4, or the brick wall in Fig. 1.5. Examine such structures at your leisure, but do not be too

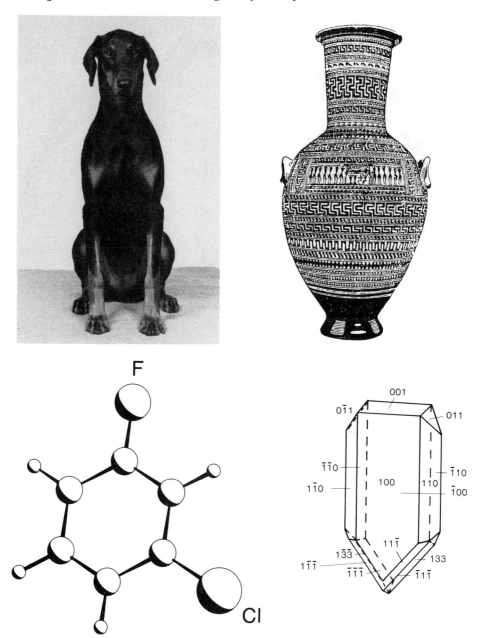

Fig. 1.2—(a) The Dobermann, Vijentor Seal of Approval at Valmara. (b) Grecian urn. (c) 3-Chlorofluorobenzene molecule. Circles in increasing order of size represent H, C, F and Cl atoms. (d) Potassium tetrathionate ($K_2S_4O_6$) crystal. The mirror symmetry relates crystal faces with Miller indices (hkl) and $(h\bar{k}l)$, the crystallographic labels for the faces.

critical about the stains on a few of the tiles, or the chip off the occasional brick. Perfect tiled floors and perfect brick walls are, like perfect crystals, conceptual.

Each of the patterns illustrated in Figs. 1.4 and 1.5 contains both a motif (a tile or a brick) and a mechanism for repetition of the motif in a regular manner. Ideally,

Fig. 1.3—Vijentor Seal of Approval at Valmara: (a) object, (b) mirror image. (From a three-dimensional point of view, there are three symmetry elements here: *m* planes through each dog and between the two of them, and a rotation axis (2-fold) along the line of intersection of these planes.)

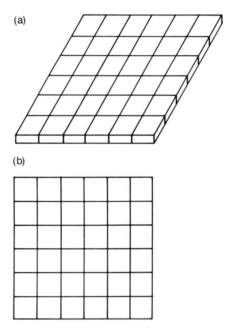

Fig. 1.4—Tiled floor: (a) parallel perspective view, (b) plane view.

Fig. 1.5—Brick wall.

the symmetry of repetition implies infinite extent, because indistinguishability between the original pattern and that found after translation must obtain. A brick wall is limited by the terminations of the building of which it is a part, just as the stacking of the units of pattern of a crystal are limited by the crystal faces. In both examples, we may utilize symmetry rules appropriate to infinite patterns provided that the number of repetitions that we are examining, by eye in the case of the wall or by X-rays in the case of the crystal, is very large compared with the size of the repeating unit itself.

Real structures, then, do not have the perfection ascribed to them by the geometrical illustrations to which we are accustomed. Nevertheless, we shall find it both important and rewarding to apply symmetry principles to objects as though they were perfect, and so build up a symmetry description of all objects and repeated patterns in terms of a small number of symmetry concepts.

1.2 DEFINING SYMMETRY

Symmetry is not an absolute property of a body that exhibits it: the result of a test for symmetry may depend upon the nature of the examining probe used. For example, the crystal structure of chromium may be represented by the body-centred cubic unit cell shown in Fig. 1.6(a). The atomic positions are those obtained from an X-ray diffraction analysis of the crystal: the atom at the centre of the unit cell is, to X-rays, identical to the atoms at the corners of the unit cell, in every unit cell of the crystal.

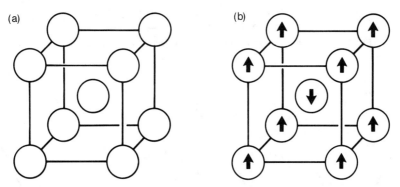

Fig. 1.6—Unit cell of the crystal structure of chromium: (a) as seen by X-rays, (b) as seen by neutrons; the arrows represent the magnetic moment vectors.

Chromium atoms have the electronic configuration $(1s)^2(2s)^2(2p)^6(3s)^2(3p)^6(3d)^5(4s)^1$, so that it is paramagnetic. If a crystal of chromium is examined by neutron diffraction techniques, we find the same positional distribution for the atoms, but their magnetic moments are not parallel. The diffraction of neutrons arises both from the atomic nuclei and from magnetic interactions with the unpaired electrons of the atoms. As a consequence, the magnetic structure is that shown by Fig. 1.6(b), where the arrows represent the directions of the magnetic moment vectors of the chromium atoms. Clearly, the symmetry† under examination by X-ray diffraction can be different from that under examination by neutron diffraction.

---

† We may note, and it will be appreciated more fully later, that elemental chromium does not have a body-centred unit cell under neutron diffraction.

An interesting example where a true symmetry is not immediately apparent arises in the X-ray diffraction examination of potassium chloride. We know that this substance has the same structure type as sodium chloride (Fig. 1.7). Fig. 1.8 shows the structure projected on to a face of the cubic unit cell. As long as we can label the atoms K^+ and Cl^-, we have no difficulty in distinguishing between them. However, in respect of X-ray diffraction, atoms are labelled by the number of electrons that they contain. The species K^+ and Cl^- are isoelectronic and, in response to moderately low resolution X-ray measurements, the unit cell has a pseudo repeat from K^+ to Cl^- corresponding to a unit of pattern having one-eighth of the volume of the true repeating unit.

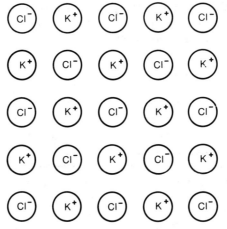

Fig. 1.7—Potassium chloride structure in projection on to a cube face, (100).

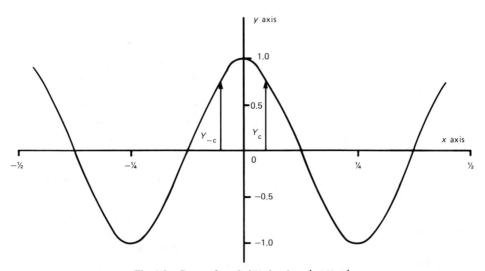

Fig. 1.8—Curve of $\cos 2\pi kX$: $k = 2$; $-\frac{1}{2} \leqslant X \leqslant \frac{1}{2}$.

In this book, we shall consider the symmetry of the positional distribution of the parts of any body, pattern, molecule or crystal, as revealed by visual inspection, microscopic examination with visible light or by X-ray methods. We shall take as our definition of symmetry *that spatial property of a body (or pattern) by which the body (or pattern) can be brought from an initial position to another, indistinguishable position, judged in terms of the above methods, by means of a certain geometrical operation—a symmetry operation.*

1.3 SYMMETRY IN SCIENCE

The manifestations of symmetry can be observed in many areas of science and, indeed, throughout nature: they are not confined to the study of molecules and crystals. In botany, for example, the symmetry inherent in the structures of flowers and reproductive systems is used as a means of classifying plants, and thus plays a fundamental role in plant taxonomy. In chemistry, symmetry occurs in individual atoms and molecules, and in crystals. Curiously, however, although crystals are only ever found to exhibit one-, two-, three-, four- or six-fold symmetry, flowers with five- or seven-fold symmetry are abundant, and even higher symmetry occurs; examples of five- and seven-fold symmetry in molecules are known. The reasons for the limitation on symmetry in crystals will emerge later.

Symmetry arises also in mathematics and physics. Consider the equation

$$X^4 = 16 \tag{1.1}$$

The roots of (1.1) are $X = \pm 2$ and $X = \pm 2\mathbf{i}$, and we can see immediately that these solutions have a symmetrical distribution. The differential equation

$$\mathbf{d}^2 Y/\mathbf{d}X^2 + k^2 Y = 0 \tag{1.2}$$

represents a type encountered, for example, in the solution of the Schrödinger equation for the hydrogen atom. The general solution of (1.2) is

$$Y = A \exp(\mathbf{i}kX) + B \exp(-\mathbf{i}kX) \tag{1.3}$$

where A, B and k are constants. If we consider a reflexion symmetry that converts X into $-X$, then the solution would become

$$Y = A \exp(-\mathbf{i}kX) + B \exp(\mathbf{i}kX) \tag{1.4}$$

Differentiating (1.4) twice with respect to X shows that this equation is a solution of (1.2). If, instead of the reflexion symmetry, we apply a translation symmetry that transforms X into $X + t$, where t is a constant, we find again that the equation and its solution have similar symmetry properties.

A single-valued, continuous, periodic function, defined between the limits $X = -\frac{1}{2}$ and $X = \frac{1}{2}$, can be represented by a series of cosine and sine terms known as a Fourier series:

$$Y = A_0 + 2 \sum_{k=1}^{k=\infty} A \cos 2\pi kX + B \sin 2\pi kX \tag{1.5}$$

A typical cosine term, shown in Fig. 1.8 as a wave, exhibits reflexion symmetry $(Y_c = Y_{-c})$ across the line $X = 0$, and is called an even function. In contrast, a typical sine term (Fig. 1.9) shows rotational symmetry $(Y_s = -Y_{-s})$ about the origin $(X = Y = 0)$, and is called an odd function.

Finally in this section, consider a cube constructed from twelve $1\,\Omega$ resistors, as shown in Fig. 1.10. Let an electrical circuit be completed through the points A and G, which lie on a symmetry line in the cube; the planes through $ACEG$, $ADFG$ and

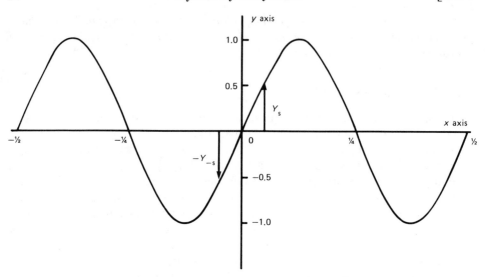

Fig. 1.9—Curve of $\sin 2\pi kX$: $k = 2$; $-\frac{1}{2} \leqslant X \leqslant \frac{1}{2}$.

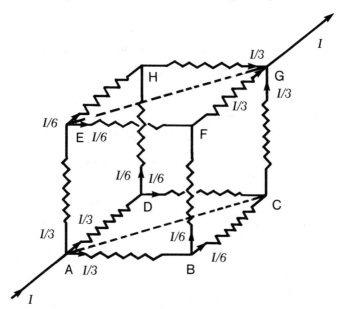

Fig. 1.10—Cubic network of 1 Ω resistors.

ABGH are all *m* planes, of the type that we have already discussed. We can use the symmetry properties of the circuit to determine the resistance of any path from *A* to *G*. The symmetry equivalence of the three paths from *A* and the three to *G* require that the currents in *AB*, *AD*, *AE*, *CG*, *FG* and *HG* are all equal to *I*/3, flowing in the directions shown by the arrows. The *m* symmetry requires that the currents in *EF* and *EH* are equal; thus, each is *I*/6. Similarly, the currents in *BC* and *DC* are also *I*/6 each, from which it follows that those in *BF* and *DH* are each *I*/6. Thus, any path through the cube from *A* to *G* has a resistance of 5/6 Ω.

Symmetry, then, can be seen to be a feature of both everyday and scientific life. In the following chapters, we shall develop its applications to the study of crystals and molecules, regarding each first as a non-repeating, one-off body. The description of the symmetry of such an object is given in terms of a point group, in which all its symmetry operations concur in one point—and this is applicable to both two- and three-dimensional objects, as we shall see.

Later, we shall have to consider, especially in relation to the structures of crystals at the atomic level, the interplay of symmetry and repetition. This gives rise in two dimensions to the plane groups, and in three dimensions to the space groups. Finally, we shall consider the elements of group theory, a generalized theory stemming from the simpler geometrical examples, and we shall consider, from a number of examples, how it may be applied to a wider variety of physical and chemical situations.

However, before embarking on these topics, we shall have to spend some time, in Chapter 2, sharpening our notions in terms of crystals, and in acquiring the requisite descriptive tools.

1.4 SYMMETRY NOTATION

Unfortunately, there exist two separate notations for symmetry, both of which are in common use. Crystallographers use the Hermann–Mauguin notation, and spectroscopists and theoreticians favour the Schoenflies notation. The main reason for this situation is largely traditional, but once one has become familiar with symmetry, its notations present little problem. However, for those studying symmetry for the first time, and perhaps in two notations at the same time, something better could and should be done.

In Appendix 2, an attempt has been made to produce a single symmetry notation based, for good reasons, on the Hermann–Mauguin symbols. It is not employed in the main text because it is not a standard notation, although the differences from the standard notation are small. In the ensuing chapters, we shall indicate the use of both the Hermann–Mauguin and the Schoenflies notations. We shall introduce symmetry symbols as we progress with the study of point groups and space groups, rather than by an extensive tabulation of symbols at an early stage.

PROBLEMS 1

1. Find the following objects in the home, or elsewhere, and study their symmetry. Report the numbers and nature of symmetry axes and/or symmetry planes present.
   (a) Cup
   (b) Rectangular table
   (c) Glass tumbler
   (d) Inner tray of matchbox
   (e) Outer sleeve of matchbox, ignoring the label
   (f) Brick
   (g) Round pencil (unsharpened)
   (h) Round pencil (sharpened conically)
   (i) Chair
   (j) Gaming die

2. Study the following patterns. Illustrate each pattern by the minimum number of representative points—where each such point is, say, the top left-hand corner of the pattern motif.
   (a) Tiled floor
   (b) Brick wall
3. Twelve $1\,\Omega$ resistors are connected so as to form the outline of a regular octahedron (same symmetry as a cube). A battery is connected across a pair of opposite apices of the octahedron. Use the symmetry of the octahedron to determine the resistance to the given current path.
4. State whether the following functions of X have even or odd symmetry.
   (a) X^3
   (b) $\sin^2 X$
   (c) $\exp(-X)$
   (d) $X^{-1} \sin X$
   (e) $X^2 - X$
   (f) $X \cos X$
5. Write in capitals those letters of the alphabet that cannot exhibit symmetry; your answer may depend on how you form the letters.

2

Geometry of Crystals

And there is a certain facility for learning all
other subjects in which we know that those who
have studied geometry lead the field.

Plato, *circa* 380 BC: *The Republic*, Book VII

2.1 INTRODUCTION

Crystallography, the science of structure, had its origins in the field of mineralogy, and involved the recognition, description and classification of naturally occurring substances. Nowadays, crystallography is a subject of vast extent, encompassing the study of structures from those of the simplest elemental substances to those of proteins, enzymes and polycrystalline aggregates of all types.

The distinctive feature of crystals is that they are bounded by flat surfaces which intersect in straight lines; nothing else in Nature exhibits such simple outlines. Secondly, measurements show that the interfacial angles (measured between the normals to adjacent faces, see Section 2.6.1) are constant for the same orientations in different crystals of the same substance, regardless of the areas of the bounding faces. The areas depend upon the conditions of growth, and thus can be accidental. The symmetry exhibited by a substance must be sought in the set of face normals, so don't be misled by any lack of symmetry in the areas of faces, in any practical situation. To assist recognition of symmetry, the crystals depicted in this book are idealized, that is, the areas and shapes of symmetry-related faces have been drawn so as to match one another, as appropriate.

2.2 REFERENCE AXES

In order to link the various observed interfacial angles into a coherent spatial arrangement, or aggregate, we need to refer all the crystal faces (or face normals) to a common set of reference axes. In some instances, the rectangular axes of coordinate geometry are appropriate, so we shall begin our discussion with them. But for many crystals a set of oblique (non-rectangular) axes will turn out to be appropriate, so we shall have to generalize as soon as possible.

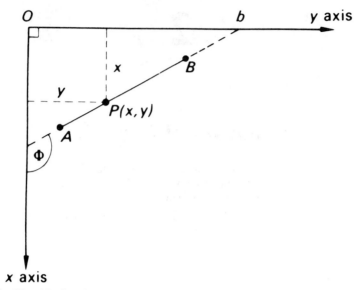

Fig. 2.1—Line *AB* referred to rectangular axes. In this and similar diagrams, the axes have been set this way round in order to conform with the standard orientation of projections of three-dimensional information. For example, see the illustrations of stereograms (Fig. 3.11 ff) and of space groups (Fig. 5.10 ff).

In two dimensions, a straight line may be referred to rectangular axes, and described by an equation. Thus, the line *AB* (Fig. 2.1) may be represented by

$$Y = mX + b \tag{2.1}$$

where *m* (=tan Φ) is the *slope* of the line, and *b* is its *intercept* on the *y* axis, that is, the value of *Y* at *X* = 0. Any point *P*(*X*, *Y*) on the line satisfies equation (2.1).

The line *AB* could be referred to oblique axes (Fig. 2.2); its equations would then be

$$Y = MX + b \tag{2.2}$$

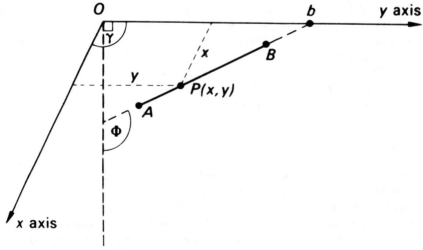

Fig. 2.2—Line *AB* referred to oblique axes.

where $M = (\tan \Phi \sin \gamma - \cos \gamma)$ and b has the same definition as before. In a particular example with $\gamma = 120°$, the equation becomes

$$Y = (\sqrt{3} \tan \Phi + 1)X/2 + b \qquad (2.3)$$

Evidently, this form of the equation is not convenient when we have to refer a line to oblique axes.

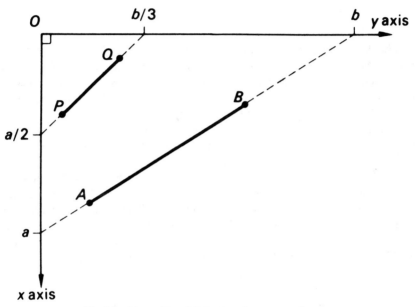

Fig. 2.3—Lines AB and PQ referred to rectangular axes.

It is possible to describe the line in other ways. In Fig. 2.3, the line AB intersects the x and y axes at a and b respectively, and has a slope m. At $X = a$, $Y = 0$, and using (2.1), we obtain

$$ma + b = 0 \qquad X/a + Y/b = 1 \qquad (2.4)$$

Equation (2.4) is the *intercept* form of the equation of the straight line AB, and will be chosen as a reference or *parametral* line. This equation, which involves the intercepts a and b on the x and y axes respectively, is applicable to any set of axes, rectangular or oblique. A second form of the equation is

$$X \cos \chi + Y \sin \chi = d \qquad (2.5)$$

where χ (equal to $\Phi - 90$) is the angle between the line AB and its normal from O, and d is the distance along the normal from O to the line. Equation (2.5) may be derived from the more general equation (2.9), by setting ω equal to zero, whereupon ψ becomes equal to $90 - \chi$; (2.5) is restricted to rectangular axes, but takes the form

$$X \cos \chi + Y \sin(\chi + \delta) = d \qquad (2.6)$$

when the axes are inclined at $(90 + \delta)$. Equations (2.5) and (2.6) can be regarded as variants of the intercept equation of the straight line.

Consider next any other line, such as PQ, and let it intercept the x axis at $a/2$ and the y axis at $b/3$. This line can be identfied by two numbers, h and k, defined such that h is the ratio of the intercept made on the x axis by the parametral line to that made by the line PQ, and k is the corresponding ratio of the intercepts made on the y axis by the same two lines. Thus,

$$h = a/(a/2) = 2 \tag{2.7}$$

and

$$k = b/(b/3) = 3 \tag{2.8}$$

PQ may be described now as the line (23)—'two-three'; it follows that the parametral line is (11). Although the values of a and b are unspecified, once the parametral line is chosen, relative to the reference axes, any other line is uniquely defined by the values of h and k determined, following (2.7) and (2.8). Note that h and k do not have to be integers, although they have been so chosen.

From the foregoing discussion, it should be clear that reference axes may be chosen in an infinity of ways (Fig. 2.4). Convention, however, dictates a choice such as in (c), where the axes are chosen to be parallel to the perimeter lines (important features) of the rectangle, rather than as in (a). If AB is chosen as the line (11), as in (c), then the lines PQ, QR, RS and SP may be designated (10), (01), ($\bar{1}$0) and (0$\bar{1}$) respectively: now, h and k are integers; a line parallel to an axis may be said to intercept it at infinity, so that for PQ, $h = a/a = 1$ and $k = b/\infty = 0$. A line that intercepts an axis on the negative side of the origin O has a corresponding negative value of h or k, so that for SP, $h = 0$ and $k = -1$, generally written as $\bar{1}$—'bar-one'

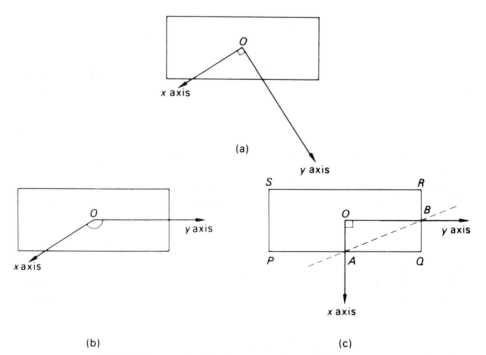

Fig. 2.4—Rectangle referred to rectangular and oblique axes.

('one-bar' in the USA). It can be verified easily that this simple set of h, k values would not be obtained with the tilted rectangular axes of Fig. 2.4(a) or with the oblique reference axes of Fig. 2.4(b).

In considering a parallelogram, however, oblique axes (Fig. 2.5(b)) *are* more appropriate. It is left as an exercise to the reader to show that, if the line AB is (11) (Fig. 2.5), then PQ, QR, RS and SP are again (10), (01), ($\bar{1}$0) and (0$\bar{1}$) respectively, only if the reference axes are chosen to be parallel to the sides of the parallelogram as in (b), rather than the rectangular axes in (a).

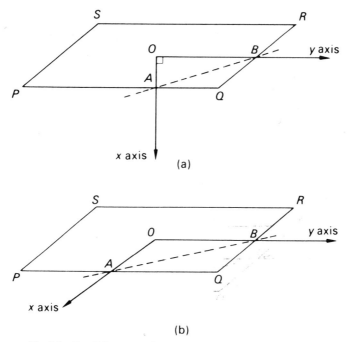

Fig. 2.5—Parallelograms referred to rectangular and oblique axes.

2.2.1 Crystallographic axes

Three reference axes are needed for the description of a crystal. An extension of the arguments already developed leads to the adoption of x, y and z axes, parallel to important directions in the crystal. For the crystal illustrated in Fig. 2.6, a conventional set of axes is shown in thick dashed lines. These axes are parallel to directions which, as we shall discuss later, are related to the symmetry of the crystal.

It is usual to work with *right-handed* axes, which will not always be rectangular. Thus, if y and z are in the plane of the paper (Fig. 2.7), x is directed forward, towards the reader; the succession $x \rightarrow y \rightarrow z \rightarrow x$ is an anticlockwise screw movement, which is one way of describing right-handed axes. The angles between pairs of axes are called α, β and γ, such that α is the angle between y and z (or 'opposite' x), β is the angle between z and x, and γ is the angle between x and y. Notice the mnemonic connexion between x, y and z, and α, β and γ.

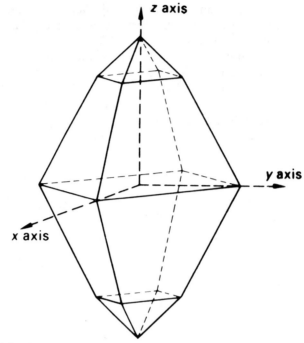

Fig. 2.6—Idealized crystal, showing the conventional x, y- and z-axes.

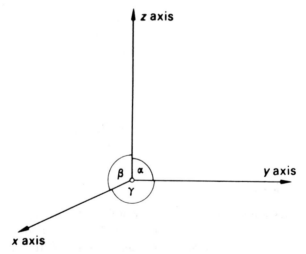

Fig. 2.7—Right-handed crystallographic axes x, y and z, and interaxial angles α, β and γ.

2.3 EQUATION OF A PLANE

The plane ABC (Fig. 2.8) intercepts the rectangular x, y and z axes at A, B and C respectively: ON is the perpendicular from the origin O to the plane; it has the length d and its direction cosines (see Appendix 4) are $\cos \chi$, $\cos \psi$ and $\cos \omega$, with respect to the x, y and z axes. A point P, with the coordinates X, Y and Z, lies in the plane

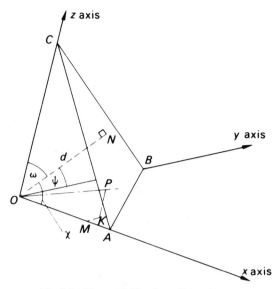

Fig. 2.8—Plane *ABC* in three dimensions.

ABC; *PK* is drawn parallel to *OC* and meets the plane *AOB* at *K*, and *KM* is drawn parallel to *OB* to meet *OA* at *M*. The lengths of *OM*, *MK* and *KP* are then *X*, *Y* and *Z* respectively. Since *ON* is the projection of *OP* on to *ON*, it is equal to the sum of the projections of *OM*, *MK* and *KP*, all on to *ON*. Hence

$$d = X \cos \chi + Y \cos \psi + Z \cos \omega \qquad (2.9)$$

In the triangle *OAN*, $d = OA \cos \chi$. Similarly, $d = OB \cos \psi = OC \cos \omega$. Thus, if $OA = a$, $OB = b$ and $OC = c$,

$$X/a + Y/b = Z/c = 1 \qquad (2.10)$$

which is the intercept equation of the plane *ABC*. Although derived for rectangular axes, this equation is applicable to all sets of axes provided that *a*, *b* and *c* are the intercepts of the parametral plane on the *x*, *y* and *z* axes respectively.

2.4 MILLER INDICES

The faces of a crystal are planes that need to be defined in three dimensions. Once the crystallographic axes are chosen, one of the faces may be defined as a parametral plane and any other plane then described in terms of three numbers, *h*, *k* and *l*; this notation was used first by Miller in 1839. It is an experimental observation that, in crystals, provided the parametral plane is assigned suitable integral values of *h*, *k* and *l*, usually (111), then the Miller indices of all other faces on the crystal are integral—generally *small* integers. This fact is contained in the Law of Rational Indices (sometimes Rational Intercepts), which is explained after the study of lattices in Chapter 4.

In Fig. 2.9, let the plane *ABC* be drawn so as to intercept the crystallographic *x*, *y* and *z* axes at *a*, *b* and *c* respectively; *ABC* is chosen as the parametral plane (111). Another plane *LMN* makes intercepts *a/h*, *b/k* and *c/l* along the *x*, *y* and *z* axes; its

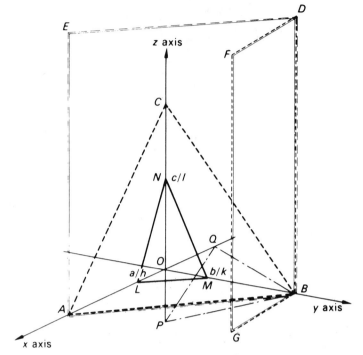

Fig. 2.9—Miller indices of crystal planes.

Miller indices are expressed by the ratios of the intercepts of the parametral plane to those of the plane LMN. If, for example, $a/h = a/4$, $b/k = b/3$ and $c/l = c/2$, then LMN is the plane (432). If fractions remain after the division, they are cleared by multiplying throughout by the lowest common denominator.

A plane such as $ABDE$ is parallel to the z axis; its intercept on this axis is infinite and its Miller indices become (110). Similarly, the plane $BDFG$ is (010). Miller indices can be negative: a plane that makes the intercepts $-a/2$, b and $-c/3$ is designated ($\bar{2}1\bar{3}$). Thus, the indices (hkl) specify a plane unambiguously in terms of the parametral plane: no assignment of values to a, b and c is needed; they are defined implicitly by the choice of parametral plane. It may be noted that any plane that intercepts all three crystallographic axes could be chosen as a parametral plane. Usually, one particular selection leads to the simplest values for h, k and l (generally less than 5) for other faces on the crystal. If, in Fig. 2.9, the plane LMN had been chosen as (111), then ABC would have been (346).

To summarize these results, we may say that the plane (hkl) makes intercepts a/h, b/k and c/l on the crystallographic x, y and z axes respectively, where a, b and c are the corresponding intercepts made by the parametral plane (111). From (2.10), it follows that the intercept form of the equation of the plane (hkl) may be written as

$$hX/a + kY/b + lZ/c = 1 \qquad (2.11)$$

The equation of the parallel plane passing through the origin of the axes is

$$hX/a + kY/b + lZ/c = 0 \qquad (2.12)$$

since it must satisfy the condition $X = Y = Z = 0$.

2.4.1 Miller–Bravais indices

In crystals that exhibit hexagonal symmetry (see Chapter 3), four axes of reference may be used; they are designated x, y, u and z. The x, y and u axes lie in one plane and at 120° to one another, and the z axis is normal to this plane (Fig. 2.10). As the parametral plane has symmetry-related equivalent faces at intervals of 120° around the z axis, the intercepts a, b and u' on the x, y and u axes respectively are equal in length, that is, $a = b = u'$. As a consequence, planes in crystals of hexagonal (and trigonal) symmetry may, conveniently, be described by the four Miller–Bravais indices h, k, i and l. These indices are not all independent, as is shown by the following argument.

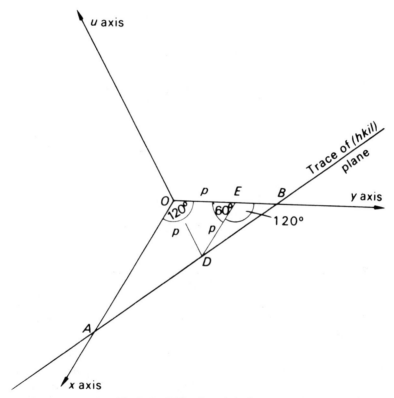

Fig. 2.10—Miller–Bravais indices.

Fig. 2.10 shows the trace of any $(hkil)$ plane intersecting the x, y and u axes at A, B and D respectively; DE is drawn parallel to AO. Since OD bisects the angle $\widehat{AOB}$, triangle ODE is equilateral. Now triangles OAB and EDB are similar, so that $EB/ED = OB/OA = (b/k)/(a/h)$. Since $EB = b/k - p$, we have $(b/k - p)/p = (b/k)/(a/h)$, or $p = ab/(ak + bh)$ which, from the equality of a and b becomes $p = a/(h + k)$. But p is the distance OD, the intercept u/i on the negative side of the u axis. Thus $i = -(h + k)$.

Then, it follows that a plane making intercepts $a/2$, $b/3$ and $c/4$ on the x, y and z axes respectively, would be labelled $(23\bar{5}4)$, and the parametral plane would be $(11\bar{2}1)$.

2.5 ZONES

Examination of most well formed crystals reveals that often faces are often symmetrically arranged in sets of two or more with respect to certain directions in the crystal. In other words, most crystals exhibit symmetry; this property is an external manifestation of the ordered arrangement of atoms and molecules of which the crystal is composed. Fig. 2.11 illustrates zircon, $ZrSiO_4$, a very symmetrical crystal.

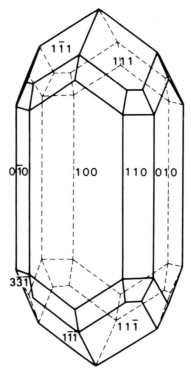

Fig. 2.11—Zircon ($ZrSiO_4$) crystal.

It is evident that several faces on this crystal have a given direction in common. Such faces are said to lie in one and the same *zone*, and the common direction is called the *zone axis*. In the illustration of zircon, it may be seen that all the vertical faces lie in one zone, for which the zone axis is parallel to the vertical direction in the crystal: all normals to these faces are perpendicular to the Z axis. Several other sets of parallel edges (zones) are also present on this crystal.

A zone can be defined by two intersecting faces. Their line of intersection, extending the faces if necessary, defines the zone axis. Thus, if $(h_1k_1l_1)$ and $(h_2k_2l_2)$ are two planes, their line of intersection, passing through the origin, is given by the solution of the equations, following (2.12),

$$h_1X/a + k_1Y/b + l_1Z/c = 0 \qquad\qquad (2.13)$$

and

$$h_2X/a + k_2Y/b + l_2Z/c = 0 \qquad\qquad (2.14)$$

that is, by the straight line

$$\frac{X}{a(k_1 l_2 - k_2 l_1)} = \frac{Y}{b(l_1 h_2 - l_2 h_1)} = \frac{Z}{c(h_1 k_2 - h_2 k_1)} \tag{2.15}$$

which may be written as

$$X/aU = Y/bV = Z/cW \tag{2.16}$$

where $[U\,VW]$ is the *zone symbol*. This result applies to any set of reference axes. A simple evaluation of U, V and W consists in writing down the indices of both planes twice, disregarding the first and last columns and cross-multiplying the remainder; the upward direction is given a positive sign, and the downward direction a negative sign, as with the evaluation of determinants. Here is an example: two planes are (123) and $(2\bar{3}1)$. What is their zone symbol? Following the rule just given:

$$1\;\begin{vmatrix} 2 & 3 & 1 & 2 \end{vmatrix}\;3$$
$$2\;\begin{vmatrix} \bar{3} & 1 & 2 & \bar{3} \end{vmatrix}\;1$$

$U = 2 - (-9) = 11;\ V = 6 - 1 = 5;\ W = -3 - 4 = -7$, and the zone symbol is $[11, 5\bar{7}]$. It may be noted that a two-digit index is followed by a comma.

2.5.1 Weiss zone equation

If $[UVW]$ represents the zone defined by $(h_1 k_1 l_1)$ and $(h_2 k_2 l_2)$, any other face lying in the same zone must satisfy the equation

$$hU + kV + lW = 0 \tag{2.17}$$

This equation is the Weiss zone equation, sometimes called the Weiss zone law, and it may be derived by the following argument. The equation of the plane parallel to (hkl) and passing through the origin is, from (2.12),

$$hX/a + kY/b + lZ/c = 0 \tag{2.18}$$

If $(h_1 k_1 l_1)$, $(h_2 k_2 l_2)$ and (hkl) are cozonal, (2.18) must, from (2.16), contain the line

$$X/aU = Y/bV = Z/cW \tag{2.19}$$

Substitution of (2.19) in (2.18) leads directly to (2.17).

The directions given by two zone symbols $[U_1 V_1 W_1]$ and $[U_2 V_2 W_2]$ define a face (hkl) common to both zones. From (2.17)

$$hU_1 + kV_1 + lW_1 = 0 \tag{2.20}$$

and

$$hU_2 + kV_2 + lW_2 = 0 \tag{2.21}$$

These equations may be solved for the ratio $h:k:l$ in a manner similar to the solution of (2.13) and (2.14) for the ratio of $U:V:W$. Thus,

$$h = V_1 W_2 - V_2 W_1 \tag{2.22}$$

$$k = W_1 U_2 - W_2 U_1 \tag{2.23}$$

$$l = U_1 V_2 - U_2 V_1 \tag{2.24}$$

Thus, if two zone symbols are [13̄2] and [02̄1̄], the face common to both zones is (1̄12). It may be noted that if the indices in the cross-multiplication scheme are written down in the reverse order of rows, then the symbol (11̄2̄) is obtained, which represents the plane, parallel to (1̄12), on the other side of the origin. Similarly, the zone symbols [*UVW*] and [*ŪV̄W̄*] represent coincident lines, but with opposite directions.

It should be evident that a study of the external features, or *morphology*, of a crystal does not permit one to distinguish between, for example, (123) and (246). In a morphological description of a crystal, however, such a distinction is not required, and the simpler indices are always used. If the reader is studying also the diffraction of X-rays from crystals, the use of general indices (*nh*, *nk*, *nl*) will be encountered; they represent, in that context, a plane (strictly, a *family* of parallel equidistant planes), parallel to (*hkl*), but $1/n$ times the perpendicular distance of (*hkl*) from the origin (Fig. 2.12).

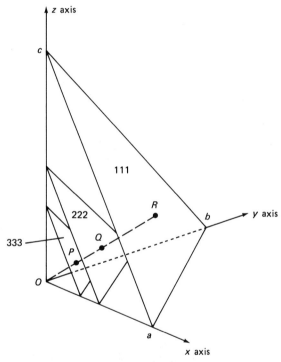

Fig. 2.12—General indices (*nh*, *nk*, *nl*); *OR* is perpendicular to the three planes shown, with $OP = OR/3$ and $OQ = OR/2$.

2.6 PROJECTION OF THREE-DIMENSIONAL FEATURES

The analytical descriptions of planes and zones given so far is inadequate for the simultaneous presentation of the symmetry revealed by the faces of a crystal (or by their normals). It is important to be able to represent a crystal in a two-dimensional drawing while preserving the fundamentally important interfacial angles truthfully.

In projecting the information contained in the interfacial angles of crystals, a method is needed which preserves angular truth: two such projections are the gnomonic projection and the stereographic projection. For our purposes, the stereographic projection is the more useful, and it will be described next.

2.6.1 Stereographic projection

This method of projection will be described with reference to the idealized crystal shown in Fig. 2.13. It shows three distinct sets of symmetry-related faces. They are six faces derived from the cube, eight from the octahedron and twelve from the rhombic dodecahedron. (Each face on the crystal drawing has a related parallel face on the actual crystal; for example, d (shown) is related to d' on the opposite side, across the origin of the axes at the centre of the crystal.) Each such set of faces is called a *form*, and the term may be applied to a set of symmetrically related lines, such as zone axes, as well as to planes.

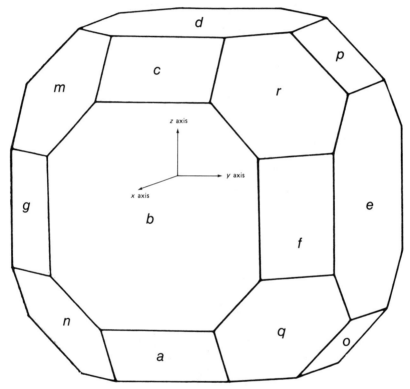

Fig. 2.13—Cubic crystal showing three forms of planes: cube—b, e and d, and parallel faces; octahedron—m, n, q and r, and parallel faces; rhombic dodecahedron—g, a, f, c, o and p, and parallel faces.

From a point, for convenience the centre of the crystal, lines are drawn perpendicular to all of the faces of the crystal. A sphere of arbitrary radius is described about the bundle of radiating normals, its centre being the point from which the normals radiate. The normals are then produced, if necessary, so as to intersect the surface of the sphere (Fig. 2.14). This construction produces a spherical projection of the crystal.

In Fig. 2.15, the plane of projection is marked $ABCD$, and it intersects the sphere in the *primitive circle*. The portion of the plane of projection enclosed by the primitive circle is the primitive plane, or just the *primitive*.

Next, the point of intersection of each normal with the upper hemisphere, e.g. R, including the primitive circle, is joined to the lowest point P on the sphere. The

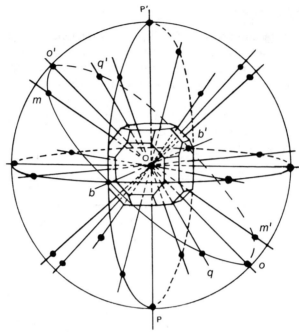

Fig. 2.14—Spherical projection of the crystal in Fig. 2.13.

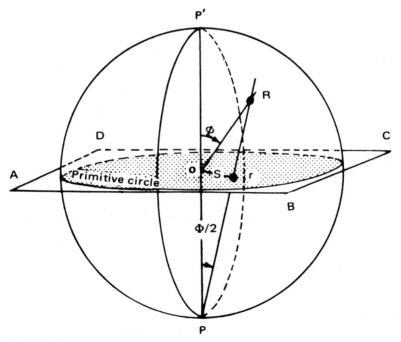

Fig. 2.15—Development of the stereographic projection from the spherical projection.

intersection of each such line with the primitive represents the stereographic projection, or *pole*, of the corresponding face of the crystal. On the stereographic projection, or *stereogram*, it is significant by a dot, e.g. *r*.

If the crystal is oriented with respect to the sphere such that the vertical direction *d′d* in the crystal is along the diameter *PP′* of the sphere, then the normals to the faces *g*, *b*, *f*, *e*, *g′*, *b′*, *f′* and *e′*, for which *dd′* is the zone axis, lie in the plane of projection and intersect the sphere in the primitive circle. Clearly, lines drawn from *P* to points of intersection on the lower hemisphere will meet the plane of projection outside the primitive circle. In order to avoid increasing the size of the stereogram unduly, such intersections are joined to the upper point *P′* on the sphere, and the corresponding poles are shown by open circles. Fig. 2.16 shows the zone *m*, *b*, *q*, *o*, *m′*, *b′*, *q′*, *o′* projected both conventionally (within the primitive) and only from the lower point *P*. The unwieldly extension of the stereogram is apparent.

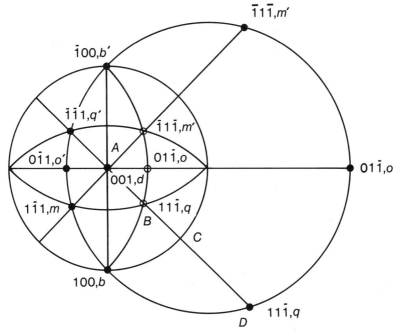

Fig. 2.16—Stereogram of the crystal of Fig. 2.13, showing the zone *m*, *b*, *q*, *o*, *m′*, *b′*, *q′*, *o′* projected both conventionally and from the lower point *P* only.

The angle $00\overline{1}\,\widehat{}\,11\overline{1}$ (the angle between the normals to (001), *d*, and (11$\overline{1}$), *d′*, is 125°16′. This angle on the sphere is represented by the distance *AD*, and also by *AC* + *CB* (90° + 35°16′). The validity of this statement is shown through Fig. 2.17. The true pole of (11$\overline{1}$) is at *D*, and the corresponding stereographic distance *AD* is $\rho \tan(125°16′/2)$, where ρ is the radius of the sphere, and of the stereogram. The conventional pole of (11$\overline{1}$) is at *B*; the stereographic distance that is equivalent to the angle of 125°16′ is *AC* + *CB*, because $00\overline{1}\,\widehat{}\,11\overline{1} = 90° + \widehat{C}\,11\overline{1}$. But $\widehat{C}\,11\overline{1} = 90° - \widehat{P}\,11\overline{1}$ and $\widehat{P}\,\overline{1}1\overline{1} = AB$, whence $\widehat{C}\,\overline{1}1\overline{1} = CB$ and, thus, $00\overline{1}\,\widehat{}\,11\overline{1} = AC + CB$ (cf. Fig. 2.16).

The stereogram is completed in Fig. 2.18. The poles should be compared via their labels with the corresponding faces on the crystal. Two important features have

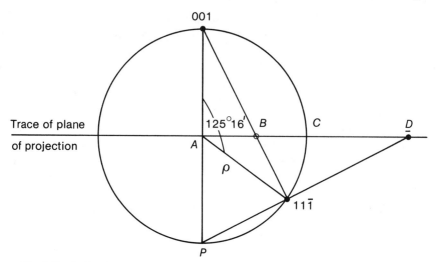

Fig. 2.17—Rationale for the conventional projection of a pole in the lower hemisphere.

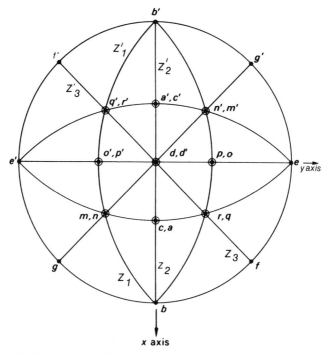

Fig. 2.18—Completed stereogram of the crystal in Fig. 2.13. The zone circles Z_1Z_1', Z_2Z_2' and Z_3Z_3' have the zone symbols [011], [010] and [1$\bar{1}$0] respectively.

emerged so far, first, that distances on a stereogram are (non-linear) measures of angles and, secondly, that all circles drawn on the sphere (Fig. 2.14) project as circles (Appendix 3). Thus, the curve Z_1Z_1' (Fig. 2.18) is an arc of a circle; it is the projection of the great circle that is inclined to the plane of projection in Fig. 2.14. A *great circle* is the trace on the sphere made by a plane which passes through the centre of the sphere. Special cases of great circles are the primitive circle, which lies in the plane of projection, and straight lines, such as Z_2Z_2', which are the projections of great circles lying normal to the plane of projection (Fig. 2.14).

The great circles on a spherical projection can be likened to the meridians on a globe of the world. Circles formed on the sphere by the intersections of planes that do not pass through the centre of the sphere are called *small circles*; they may be likened to the parallels of latitude on the globe.

The pole of a great circle is the stereographic projection of a point on the sphere that is 90° away from all points on the great circle, that is, it is the pole of a face that is at right-angles to all the faces in a given zone. Thus, the faces g, b, f, e, g', b', f' and e' all lie in one and the same zone; the face d (or d) is at right-angles to all these faces, and the pole of d (or d') is also that of the great circle (*zone circle*) under consideration.

In order to construct Fig. 2.18, the following practical principles must be understood. The interfacial angles are measured with a goniometer (Fig. 2.19) in their correct zones, whereupon the great-circle geometry, to be discussed below, is immediately applicable. Precise goniometry may be carried out on small crystals by means of an optical goniometer, an example of which is shown in Fig. 2.20.

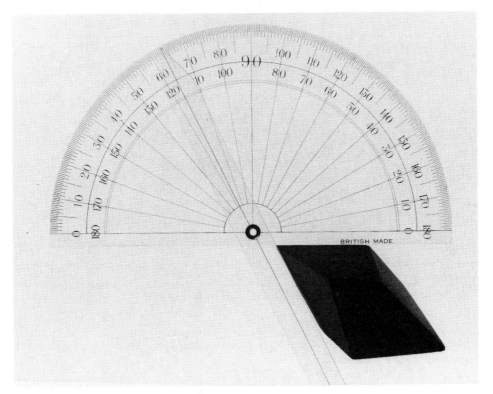

Fig. 2.19—Contact goniometer, with a crystal model in the measuring position.

Having made the necessary measurements of interfacial angles (in practice, the angles between the face normals), a crystal orientation with respect to the sphere must be selected. In the example under consideration (Fig. 2.13), we have arranged for the zone g, b, f ... to be on the primitive circle (Fig. 2.18), and for the zone b, c, d ...,

Fig. 2.20—Two-circle optical goniometer: the crystal rotates about the vertical circle, and the telescope and collimator about the horizontal circle.

normal to it, to run from bottom to top on the drawing (Z_2Z_2'). The angle $\widehat{bf}$ is found to be $45°00'$; the zone $f, r, d \ldots$, because it contains d, is normal to the plane of projection, and thus projects as the straight line Z_3Z_3', at $45°$ to Z_2Z_2'. The distance S of the pole r from the centre of the stereogram is given by (Fig. 2.21)

$$S = \rho \tan(\Phi/2) \tag{2.25}$$

where Φ is here the angle between the normals to the faces d and r. Other faces may be plotted in this manner.

A simple graphical method employing a Wulff's net (Fig. 2.22) is often sufficiently accurate. On this chart, the curves running from top to bottom are projected great circles (meridians), and those running from left to right are projected small circles (parallels). The Wulff's net in the position illustrated may be imagined to represent the global hemisphere 90°W to 90°E, with the 0°–0° line as the equator. The applications of the chart are not, of course limited to this orientation.

In using the net, its centre is pivoted about the centre of the stereogram, the interfacial angle measured along the appropriate great circle and the pole plotted.

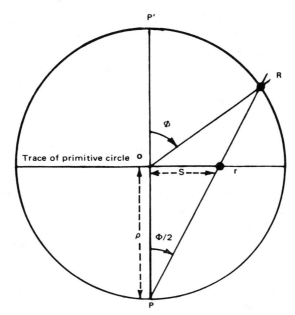

Fig. 2.21—Vertical section through *R* in Fig. 2.15.

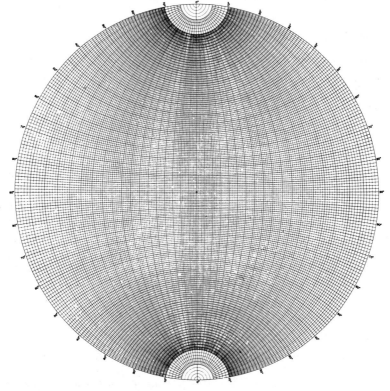

Fig. 2.22—Wulff's net.

The pole r lies at the intersection of the zones b, r, p and d, r, f, for example. Thus, if the angles $\widehat{dp}$ and $\widehat{df}$ are measured (or known), the zone circles b, r, p and d, r, f can be drawn and pole r located.

It follows from this discussion that a Wulff's net may be used to measure the angle between two poles on a stereogram. The net is aligned with the stereogram as before, and then rotated until the two poles lie on one and the same great circle; the angle is then read directly from the net. The precision of this procedure is approximately $\frac{1}{2}°$ with a net of radius about 60 mm.

The stereogram may now be indexed; this procedure consists in allocating Miller indices to each pole that has been plotted. The face r may be chosen as the parametral plane (111), because it cuts all three axes, and the remaining faces indexed, as shown in Fig. 2.23. The Weiss zone equation (2.17) may be used to confirm the accuracy of the assignments of indices. It is not necessary to write the indices for both of the poles that are represented by the ring and the dot at any position. Conventionally, the dot, which relates to the upper hemisphere, alone is indexed. Thus, if r (Fig. 2.18) is 111, then we know that q is $11\bar{1}$. The indices of the poles are written without parentheses; each triplet represents the normal hkl to the face (hkl): it should be noted that the zone symbol with the same integers $[hkl]$ does not, in general, coincide with the normal to plane (hkl).

The angular truth of the stereogram (see Appendix 3) makes it very suitable for representing many properties that depend upon angular relationships; we shall use it in connexion with interfacial angles of crystals, symmetry elements, point groups, and bond directions in molecules or complex ions.

2.6.2 Interpretation of goniometric measurements

As a further exercise in interpreting goniometric measurements, the data obtained from a crystal of Hemimorphite, $Zn_4(OH)_2Si_2O_7 . H_2O$ (Fig. 2.24), will be studied. Table 2.1 lists a selection of measured interfacial angles; it does not include all the measurements that would be made in practice.

The completed stereogram of Hemimorphite is shown in Fig. 2.25. The chosen orientation will become clear from a comparison of the stereogram and the drawing of the crystal. The indexing has been carried out with the pole s as 111: the parametral plane need not be a face developed on the crystal, but it must be a possible face, that is, it must lie at the intersection of two zones (fdf' and ere', in this case).

From the measurements, we see that the angles $100\,\widehat{\,}110$ and $001\,\widehat{\,}101$ are not equal to 45°, but that the angles α, β and γ are each 90°. The derivation of these angles from the stereogram can be understood by reference to Fig. 2.26(a) and (b).

The *axial ratios* a/b and c/b can be determined from Fig. 2.27: X, Y and Z here represent the poles 100, 010 and 001 respectively, so that $OX:OY:OZ = a:b:c$; P and Q represent the poles 110 and 011 respectively. Hence,

$$a/b = \tan \widehat{OYP} = \tan \widehat{XOP} = \tan 100\,\widehat{\,}110 \qquad (2.26)$$

and

$$c/b = \tan \widehat{OYZ} = \tan \widehat{ZOQ} = \tan 001\,\widehat{\,}011 \qquad (2.27)$$

Thus, $a/b = 0.784$ and $c/b = 0.478$ $(c/a = 0.610)$, or $a:b:c = 0.784:1.000:0.478$, as measured from the stereogram. It should be noted that the individual values of a, b

and c cannot be obtained from the stereogram, or from any other morphologial measurements.

Referring again to Fig. 2.25, if m had been chosen as 111, different indices would have arisen for certain poles: a/b and c/b would have become 1.568 and 0.956 respectively and pole h, for example, would have become 023; c/a would have remained unchanged because pole r remained as 101.

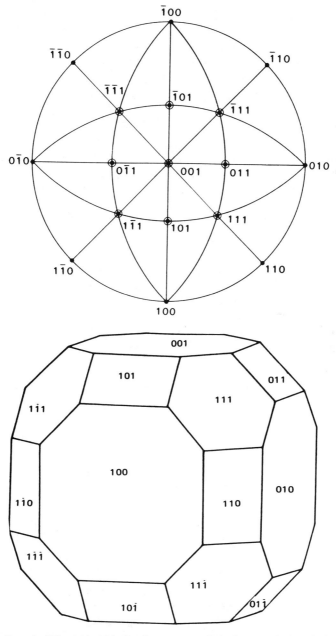

Fig. 2.23—Crystal of Fig. 2.13: (a) indexed stereogram, (b) indices attached to the faces of the crystal (see Fig. 2.13).

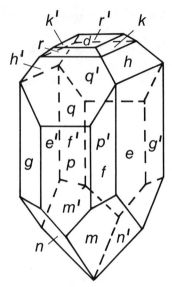

Fig. 2.24—Crystal of Hemimorphite.

Table 2.1—Gonometric results on Hemimorphite

Zone $p, q, r, d, \ldots$		Zone $e, f, p, g, \ldots$	
$\widehat{pq}$	28°43′	$\widehat{ef}$	51°51′
$\widehat{qr}$	29°53′	$\widehat{fp}$	38°06′
$\widehat{rd}$	31°25′	$\widehat{pq}$	38°03′
$\widehat{dr'}$	31°19′	$\widehat{ge'}$	51°56′
$\widehat{r'q'}$	30°03′	$\widehat{e'f'}$	51°53′
$\widehat{q'p'}$	28°35′	$\widehat{f'p'}$	38°08′
		$\widehat{pg'}$	38°01′
		$\widehat{g'e}$	52°00′

Zone $e, h, k, d, \ldots$		Zone $e, m, n, \ldots$	
$\widehat{eh}$	34°54′	$\widehat{em}$	50°48′
$\widehat{hk}$	29°36′	$\widehat{mn}$	78°22′
$\widehat{kd}$	25°34′	$\widehat{ne'}$	50°43′
$\widehat{dk'}$	25°32′	$\widehat{e'm'}$	50°51′
$\widehat{k'h'}$	29°36′	$\widehat{m'n'}$	78°25′
$\widehat{h'e'}$	34°49′	$\widehat{n'e}$	50°47′
		$\widehat{er}$	89°59′
		$\widehat{e'r'}$	90°01′

Zone $e, q, e', \ldots$		Zone $p, m, n', \ldots$	
$\widehat{eq}$	90°02′	$\widehat{pm}$	66°14′
$\widehat{qe'}$	89°58′	$\widehat{mn'}$	47°36′
$\widehat{e'q'}$	89°58′	$\widehat{n'p'}$	66°08′
$\widehat{q'e}$	90°00′	$\widehat{p'm'}$	66°16′
		$\widehat{m'n}$	47°34′
		$\widehat{np}$	66°13′

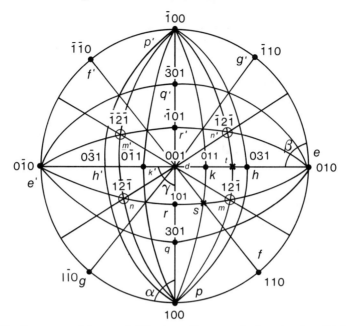

Fig. 2.25—Stereogram of the crystal of Hemimorphite; ✳ marks the poles of possible faces.

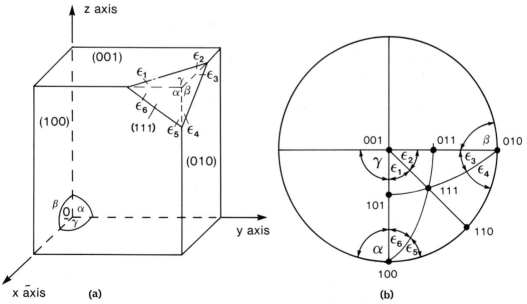

Fig. 2.26—Positions of α, β and γ on the stereogram: (a) crystal showing planes and angles (b) corresponding partial stereogram. The angle ε_1 is the angle between the edges (100)–(001) and (001)–(111), and etc. Since internormal angles are plotted on the stereogram, γ is the angle $[180 - (\varepsilon_1 + \varepsilon_2)]$, and etc.

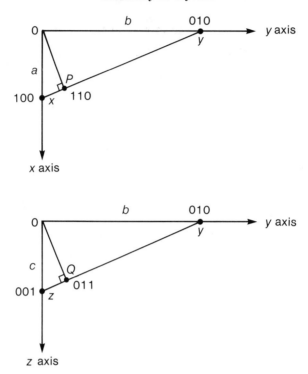

Fig. 2.27—Derivation of axial ratios from stereogram data; in some cases, the goniometric measurements may be used directly, with greater precision in the results.

The completed stereogram shows that more zone measurements could have been made than those recorded in Table 2.1. They are not concerned with any more crystal faces, but the additional data would have increased the precision of the angles plotted on the stereogram; two such zones are f, h, f' and g, m, g'.

The indexing of m as $12\bar{1}$ can now be understood fully. The great circle through p and m intersects the zone e, h, k ... in the possible pole t. The angle $00\widehat{1\;t}$ is, from the stereogram, approximately 43°30', and from the ratio c/b (0.478) and a construction like Fig. 2.27 the indices of t are $02\bar{1}$. The zones defined by the pairs of faces (100), $(02\bar{1})$ and (010), $(10\bar{1})$ are [012] and [101] respectively. Hence, m, lying at their intersection, and in the lower hemisphere, is $12\bar{1}$. This result may be checked with (2.22) to (2.24).

2.6.3 Calculations in stereographic projections

The utilization of the full accuracy of optical goniometric measurements requires the formulae of spherical trigonometry. Fig. 2.28 shows a spherical triangle ABC formed on the surface of a sphere, centre O, by the intersections of great circles represented in parts by the curves AB, AC and BC. The arcs a, b and c are the sides of the spherical triangle, and its angles will be represented by the letters A, B and C. The sides are measured by the angles that they subtend at O; thus, a is determined by the angle BOC, for example.

If A, B and C are located by the poles of three crystal faces that do not all lie in one and the same zone, then the sides a, b and c represent the corresponding interfacial

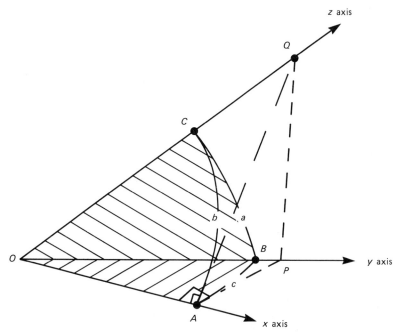

Fig. 2.28—Spherical triangle ABC; APQ is the tangent plane to the spherical surface at A.

angles. The triangle formed by 100, 0$\bar{1}$0 and 001 in Fig. 2.25 is another example of a spherical triangle, but in projection. The angle α is measured by the angle between the sides 001, 100 and 100, 0$\bar{1}$0, so that where this angle is not 90°, the obtuse angle will be determined (by convention); in the present example, α is 90°. Similar results hold for β and γ.

Referring again to Fig. 2.28, the angle A is measured by the angle between the tangents at A to the arcs CA and BA: it is the *dihedral* angle between the planes AOC and AOB. Since CA and BA are portions of zone circles, the angle at their intersection is the angle between the corresponding zone axes. This angle is, therefore, the angle between two edges, or possible edges, of a crystal, and is equal to the plane angle PAQ.

2.6.4 Formulae for spherical triangles

A spherical triangle is described completely by its six elements, three dihedral angles A, B, C, and three interfacial angles a, b and c (see Fig. 2.28). Of the many formulae offered by spherical trigonometry, shall consider just two, the cosine formula and the sine formula. Other relationships can be deduced from them, particularly when one of the elements has the value 90°. From the formulae, we can derive the precise axial ratios for Hemimorphite as $a:b:c = 0.7841:1.0000:0.4781$.

Cosine formulae
In the triangle APQ, Fig. 2.28,

$$PQ^2 = AP^2 + AQ^2 - 2AP\,AQ \cos A \qquad (A < 180°) \qquad (2.28)$$

In triangle OPQ

$$PQ^2 = OP^2 + OQ^2 - 2OP\,OQ\cos a \tag{2.29}$$

Hence,

$$AP^2 - OP^2 + AQ^2 - OQ^2 - 2APAQ\cos A + 2OP\,OQ\cos a = 0 \tag{2.30}$$

Since $\widehat{OAQ} = \widehat{OAP} = 90°$,

$$AO^2 = OQ^2 - AQ^2 = OP^2 - AP^2 \tag{2.31}$$

Thus,

$$AO^2 = APAQ\cos A + OP\,OQ\cos a \tag{2.32}$$

and

$$\cos a = (OA\,OA/OP\,OQ) + (AP\,AQ\cos A/OP\,OQ) \tag{2.33}$$

whence

$$\cos a = \cos b\cos c + \sin b\sin c\cos A \tag{2.34}$$

which is the cosine formula. Similar formulae can be derived for $\cos b$ and $\cos c$.

Sine formulae
From equation (2.34),

$$\cos^2 A = (\cos^2 a + \cos^2 b\cos^2 c - 2\cos a\cos b\cos c)/\sin^2 b\sin^2 c \tag{2.35}$$

whence

$\sin^2 A (= 1 - \cos^2 A)$

$$= (\sin^2 b\sin^2 c - \cos^2 a - \cos^2 b\cos^2 c + 2\cos a\cos b\cos c)/\sin^2 b\sin^2 c \tag{2.36}$$

Similarly,

$\sin^2 B = (\sin^2 c\sin^2 a - \cos^2 b - \cos^2 c\cos^2 a$

$$+ 2\cos a\cos b\cos c)/\sin^2 c\sin^2 a \tag{2.37}$$

Dividing (2.36) by (2.37) and replacing $\sin^2 a$, $\sin^2 b$ and $\sin^2 c$ by $(1 - \cos^2 a)$, $(1 - \cos^2 b)$ and $(1 - \cos^2 c)$, respectively, and rearranging,

$$\sin^2 A\sin^2 b/\sin^2 B\sin^2 a = \frac{(1 - \cos^2 a - \cos^2 b - \cos^2 c + 2\cos a\cos b\cos c)}{(1 - \cos^2 a - \cos^2 b - \cos^2 c + 2\cos a\cos b\cos c)} \tag{2.38}$$

Hence, and by analogy,

$$\sin A/\sin a = \sin B/\sin b = \sin C/\sin c \tag{2.38}$$

2.6.5 Polar spherical triangles

In Fig. 2.29, ABC is a spherical triangle. Arc $B'C'$ is, at all points on it, 90° from A, so that A is the *pole* of the great circle represented by $B'C'$. Similarly, B and C are the poles of $A'C'$ and $A'B'$ respectively. Triangle $A'B'C'$ is the polar triangle of ABC.

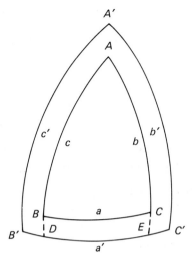

Fig. 2.29—Spherical triangle ABC and its polar spherical triangle $A'B'C'$.

Since A is the pole of $B'C'$ and C is the pole of $A'B'$, it follows that B' is 90° from the great circle arc AC. Similarly, A' and C' are 90° from BC and AB respectively. Thus, triangle ABC is the polar triangle of $A'B'C'$.

Next, let the arcs AB and AC be extended to cut $B'C'$ in D and E respectively. Now A is the pole of DE, so that DE is the measure of the angle A. But $B'E + C'D = B'C' + DE$ and, since B' and C' are the poles of CE and BD respectively, $B'E = C'D = 90°$. Thus, $B'C' + DE = a' + A = 180°$, or

$$a' = 180 - A \qquad (2.39)$$

and because triangles ABC and $A'B'C'$ are polar to each other

$$a = 180 - A' \qquad (2.40)$$

Similarly

$$b' = 180 - B \qquad (2.41)$$

$$b = 180 - B' \qquad (2.42)$$

$$c' = 180 - C \qquad (2.43)$$

$$c = 180 - C' \qquad (2.44)$$

Substituting for a', b', c' and A' and using (2.34) for triangle $A'B'C'$, we obtain

$$\cos(180 - A) = \cos (180 - B) \cos (180 - C)$$
$$+ \sin (180 - B) \sin (180 - C) \cos (180 - a) \qquad (2.45)$$

which may be simplified and rearranged to give

$$\cos a = \frac{\cos A + \cos B \cos C}{\sin B \sin C} \qquad (2.46)$$

Similarly

$$\cos b = \frac{\cos B + \cos A \cos C}{\sin A \sin C} \qquad (2.47)$$

$$\cos c = \frac{\cos C + \cos A \cos B}{\sin A \cos B} \qquad (2.48)$$

PROBLEMS 2

1. Write down the Miller indices of the planes that make intercepts as given below:

On the x axis	On the y axis	On the z axis
(a) $a/3$	$-b$	parallel
(b) $-a$	$3b/2$	$c/4$
(c) $a/4$	parallel	$-c/3$
(d) $2a$	$-b/2$	c
(e) parallel	$b/4$	$-2c$
(f) $3a$	$-3b$	$c/2$

2. Evaluate the zone symbol for each of the following pairs of planes:

(a) (001), (111)
(b) $(0\bar{1}3)$, (100)
(c) $(1\bar{2}1)$, $(\bar{3}21)$
(d) $(0\bar{1}0)$, $(\bar{1}\bar{1}3)$
(e) $(\bar{1}02)$, $(0\bar{1}3)$
(f) (123), $(\bar{2}31)$

3. In which of the zones, evaluated in 2, does each of the following faces lie? Some faces may lie in more than one zone.

(a) $(\bar{1}\bar{1}2)$
(b) $(11\bar{3})$
(c) $(0\bar{2}1)$
(d) $(01\bar{3})$
(e) $(\bar{1}1\bar{1})$
(f) (101)

4. What faces lie at the intersections of each of the following pairs of zones?

(a) $[001]$, $[111]$
(b) $[120]$, $[021]$
(c) $[113]$, $[214]$
(d) $[010]$, $[1\bar{1}\bar{1}]$
(e) $[103]$, $[100]$
(f) $[\bar{1}10]$, $[13\bar{2}]$

5. Show how the Weiss zone equation may be used to specify two possible planes in the zone $[1\bar{2}3]$. There is more than one answer.

6. If a zone is defined by the planes $(h_1k_1l_1)$ and $(h_2k_2l_2)$, show that the plane $(mh_1 + nh_2, mk_1 + nk_2, ml_1 + nl_2)$ is cozonal; m and n are integers.

7. If the normals to two planes $(h_1k_1l_1)$ and $(h_2k_2l_2)$ in an orthogonal† crystal have the direction cosines χ_1, ψ_1, ω_1 and χ_2, ψ_2, ω_2 respectively, show that the angle ϕ between the normals is given by

$$\cos \phi = \frac{(h_1h_2/a^2) + (k_1k_2/b^2) + (l_1l_2/c^2)}{[(h_1^2/a^2) + (k_1^2/b^2) + (l_1^2/c^2)]^{1/2}[(h_2^2/a^2) + (k_2^2/b^2) + (l_2^2/c^2)]^{1/2}}$$

Hence, evaluate the angle $00\overline{1}\,\widehat{}\,111$ in a cube $(a:b:c = 1:1:1)$.

8. A portion of the stereogram of an orthogonal† crystal is shown in Fig. P2.1. The axial ratios are $a:b:c = 0.5287:1.000:0.9539$, and $00\overline{1}\,\widehat{}\,111 = 63.90°$. Find, by spherical trigonometry, the angle $11\overline{1}\,\widehat{}\,102$.

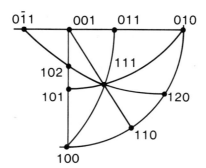

Fig. P2.1—Portion of the stereogram of an orthorhombic crystal.

9. The following goniometric measurements were recorded for a crystal of Anatase, TiO_2 (Fig. P2.2). Each datum represents the mean of eight measurements, except

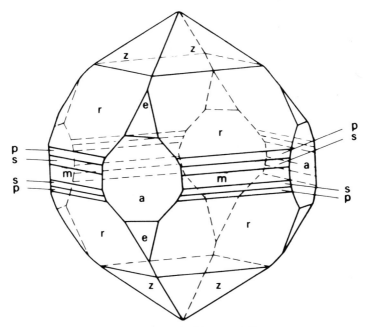

Fig. P2.2—Crystal of Anatase, TiO_2.

† Here, orthogonal means referred to mutually perpendicular axes.

that only four such measurements were made for the zone a, m, a:

Zone $a, m, a, \ldots$		Zone $m, s, p, \ldots$		Zone $a, r, r, \ldots$	
$\widehat{am}$	45°00′	$\widehat{ms}$	11°14′	$\widehat{ar}$	53°52′
$\widehat{aa}$	90°00′	$\widehat{sp}$	10°24′	$\widehat{rr}$	72°15′
		$\widehat{pr}$	11°52′		
Zone $a, s, s, \ldots$		$\widehat{rz}$	16°31′	Zone $a, e, a, \ldots$	
$\widehat{as}$	46°06′	$\widehat{zz}$	79°53′	$\widehat{ae}$	29°22′
$\widehat{ss}$	87°48′			$\widehat{ee}$	121°15′
Zone $a, z, z, \ldots$		Zone $a, p, p, \ldots$		Zone $m, e, e, \ldots$	
$\widehat{az}$	63°00′	$\widehat{ap}$	48°55′	$\widehat{me}$	51°56′
$\widehat{zz}$	54°00′	$\widehat{pp}$	82°10′	$\widehat{ee}$	76°07′

(a) Plot a stereogram, of radius equal to that of your Wulff's net, to represent the goniometric results. Let the zone a, m, a, ... be the primitive circle, and the zone a, e, a, ... run from bottom to top of your diagram.

(b) Index the poles, taking p as 111.

(c) Determine the axial ratios.

(d) Measure the angle $\widehat{pe}$ on the stereogram, and then calculate the same angle by solving the appropriate spherical triangle.

3

Point-group Symmetry

We proceeded straight from plane geometry
to solid bodies in motion without considering
solid bodies first on their own. The right
thing is to proceed from second dimension
to third, which brings us to cubes and other
three-dimensional figures.

Plato, *circa* 380 BC: *The Republic*, Book VII

3.1 INTRODUCTION

In the first chapter, we took note of the reflexion plane symmetry that certain objects may be said to possess, and we discussed the meaning of symmetry briefly. While studying zones (Section 2.5), we saw that crystals usually exhibit symmetry to some degree. In this chapter, we shall consider the symmetry of finite objects, particularly crystals and molecules.

3.2 SYMMETRY ELEMENTS AND SYMMETRY OPERATIONS

A *symmetry element* is a geometrical entity (point, line or plane) of a body, or assemblage, with which is associated an appropriate symmetry operation. The symmetry element is strictly conceptual, but it is convenient to accord it a sense of reality. The symmetry element connects all parts of the assemblage into a number of symmetrically related sets. The term *assemblage* is often useful as it describes more obviously a bundle of radiating normals, to which these symmetry concepts equally apply.

The corresponding *symmetry operation* when applied to the body converts it to a result that is indistinguishable from its initial state, and thus *reveals* the symmetry inherent in the body. In numerous cases, different symmetry operations can reveal one and the same symmetry element. Thus, 3 $(=3^1)$, 3^2 and 3^3 $(=3)$ may all be regarded as either multiple steps of 3 or single-step symmetry operations in their own right: but all are contained within the symmetry of the assemblage to which they refer, and are associated with the same single symmetry element (3). We shall consider this latter idea again in study Group Theory (Chapter 7).

3.3 POINT GROUPS

Symmetry elements may occur singly in a body, as in Fig. 1.2(a)–(d), or in certain combinations, as in Fig. 1.3. A set of interacting symmetry elements in a finite body, or just one such element, is referred to as a *point group*. A point group is defined as a set of symmetry elements all of which pass through a single fixed point: this point is taken as the origin of the reference axes for the body. It follows that the symmetry operations of a point group must leave *at least* one point unmoved; in some cases it will be a line (symmetry axis) or a plane (reflexion plane).

It can be contended that, in real objects, since they are imperfect, even if only on a microscopic scale, an indistinguishable aspect can be obtained only by a rotation of 360° (or 0°); this operation is *identity*, or 'doing nothing'. For practical purposes however, the effects of most such imperfections are small and, although our discussion of symmetry will be set up in terms of ideal geometrical objects, the extension of the results to real situations will be found to be scientifically rewarding.

We noted in Section 1.2 that the observable symmetry may depend upon the nature of the examining technique. Here, we are concerned with the symmetry shown by *directions in space*, such as the normals to the faces of a crystal or the bond directions in a molecule. Such angular relationship can be presented on stereograms (Section 2.6.1), and we shall draw fully on this method of representation in the ensuing discussion.

Several concepts in symmetry can be introduced with two-dimensional objects; subsequently, the third dimension can be introduced mainly as a geometrical extension of the two-dimensional ideas.

3.4 SYMMETRY IN TWO DIMENSIONS

Symmetry in two dimensions implies operations on an object in two-dimensional space—the plane of the paper. Any operation that would move the object out of its plane, even though indistinguishability be achieved before and after the operation, is not relevant to the two-dimensional case.

There are two types of symmetry operatons in two dimensions, those of rotation about a point and reflexion across a line.

3.4.1 Rotational symmetry

If an object can be brought into indistinguishable positions R times for a rotation of 360° about a point within its own plane, it is said to possess R-fold rotational symmetry about that point. Alternatively, we may say that an R-fold rotation brings the object into an indistinguishable position for every angular increment of $(360/R)°$. In general, R may take any value from 1 to infinity. However, in crystals, as we shall show later, R may take only the values of 1, 2, 3, 4 and 6. Although there are some *molecules* that show other rotational symmetries, such as 5-fold symmetry in biscyclopentadienyl ruthenium $C_{10}H_{10}Ru$, or, indeed, infinity symmetry in linear molecules, we shall, for the moment and for simplicity, restrict our discussion to crystallographic symmetry.

Fig. 3.1 shows two-dimensional objects that have rotational symmetry. Alongside each object is a stereogram to represent the point group of the object, with a set of equivalent points superimposed on the stereogram. Each of the objects is made up

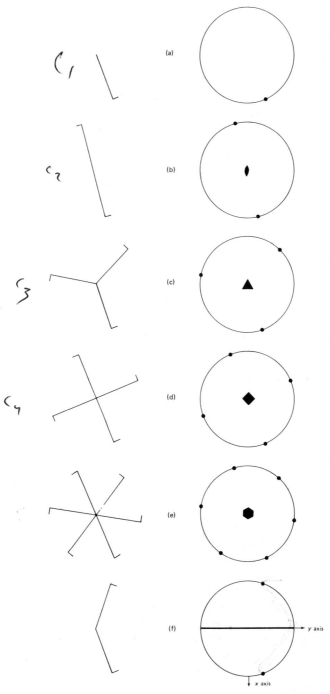

Fig. 3.1—Symmetrical two-dimensional objects and their stereograms I; because we are in two dimensions, all the poles lie on the perimeter. The graphic symbols for the symmetry elements are shown on the stereograms:

(a) Point group 1   (b) Point group 2   (c) Point group 3
(d) Point group 4   (e) Point group 6   (f) Point group m

of one or more motifs of (a), according to its rotational symmetry. The minimum number of motifs which, when operated on by the point group, builds up the whole object is called the *asymmetric unit*. In (a) the asymmetric unit is the whole object, whereas in (e) it is one-sixth of the object; we see again how a symmetry element can relate to more than one symmetry operation.

The stereograms show also the graphic symbols for the R-fold rotation axes; there is a strong mnemonic connexion between the shape of the symbol and the nature of R. The representative point of each motif is shown by ●.

3.4.2 Reflexion symmetry

In two dimensions, reflexion symmetry takes place across a line, as in Fig. 3.1(f). The line may be imagined to divide the object into two halves related to each other as is a body to its mirror image. The operation of reflexion, symbol m—graphic symbol a thick line, leads to indistinguishability before and after the operation. The operation is not one that we can perform physically on the object, unlike rotational symmetry, but we can appreciate from the object, and its stereogram, that m symmetry is present.

Reflexion symmetry involves enantiomorphous (change-of-hand) aspects in the object, whereas rotational symmetry is concerned with congruent aspects in the object. For those who are used to relating enantiomorphism to the presence of one or more chiral centres in a molecule, this description applied to the planar motifs in Figs 3.1 and 3.2 may seem inapt. However, it must be remembered that, in two dimensions, symmetry operations take place strictly in the plane. Thus, it is not possible to superimpose the two motifs in Fig. 3.1(f) by any translational movement in the plane. Accordingly, it is competent to apply the term enantiomorphism in this and similar situations.

3.4.3 Combinations of symmetry elements

The objects in Fig. 3.1(a) to (f) each contain a single symmetry element, and illustrate the two-dimensional (plane) point groups 1, 2, 3, 4 and 6. Note that even Fig. 3.1(a) which appears to have no symmetry is included, with 1-fold (identity) symmetry. Four other 'crystallographic' plane point groups consist of combinations of R and m.

Formally, five combinations may be written, namely, $1m$, $2m$, $3m$, $4m$ and $6m$; $1m$ is equivalent to m and needs no further discussion. Consider $2m$, remembering that 2 and m must intersect in a point (Fig. 3.2). We see from (a), both in the object and the stereogram, that a second mirror line is inherent in the combination $2m$; this point group may therefore be written more explicitly as $2mm$. In general, two intersecting symmetry elements imply one or more additional mutually intersecting symmetry elements. This fact can be appreciated by studying, or by redrawing, the stereograms of Fig. 3.2, starting with two elements Rm and then inserting the poles related by this symmetry.

It should be noted that in Figs 3.1 and 3.2, the poles have not been placed *on* any symmetry elements, otherwise there would have been a reduction in their number. Each stereogram thus displays, the full set of symmetry-related general poles or a *general form*, the maximum number of sites that can be generated from one pole by the point-group symmetry, leaving one pole in each initial position as well as in the

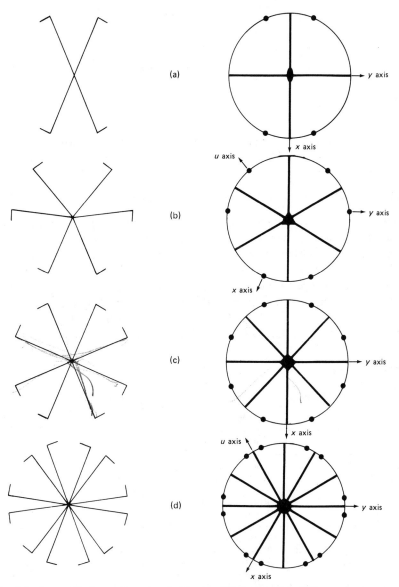

Fig. 3.2—Symmetrical two-dimensional objects and their stereograms II:
(a) 2 mm   (b) 3m   (c) 4mm   (d) 6mm

symmetry-related position after each movement. It follows, therefore, that we can only be certain of the true symmetry of an object where the general form is present on it (or its stereogram). Consider Fig. 3.3: it could be the general form of point group 2, or a *special form* of point group 2mm. Thus, the poles of a special form must lie on symmetry elements, and the entity or motif that each represents must have the symmetry of that element or a symmetry that is consistent with it—a sub-group (Section 3.4.6).

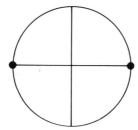

Fig. 3.3—Stereogram showing a form of two poles; with no other information, it could be either a general form in point group 2 or a special form in point group 2*mm*.

3.4.4 Two-dimensional systems and point-group notation

It is useful to refer two-dimensional objects, and their point groups, to certain two-dimensional (axial) *systems*; the system name is related to the maximum rotational symmetry present in the object. Reference axes are chosen in the closest relationship to the symmetry elements, as implied in Chapter 2. There results a precise orientational significance to the positions in the point-group symbol. This fact is a feature of the Hermann–Mauguin (crystallographic) symmetry notation that will be developed more fully under three-dimensional point groups. Table 3.1 lists the two-dimensional systems, point groups and interpretation of the symbols.

Table 3.1—Two-dimensional 'crystallographic' systems and point groups

System	Point group	Symbol meaning for each position		
		1st position§	2nd position	3rd position
Oblique	1, 2†	Rotation about a point	—	—
Rectangular	1*m*‡	Rotation about a point	$m \perp x$	—
	2*mm*	Rotation about a point	$m \perp x$	$m \perp y$
Square	4	Rotation about a point	—	—
	4*mm*	Rotation about a point	$m \perp x, y$	m 45° to x, y
Hexagonal	3, 6	Rotation about a point	—	—
	3*m*	Rotation about a point	$m \perp x, y, u$	—
	6*mm*	Rotation about a point	$m \perp x, y, u$	m along x, y, u

† 2 is the two-dimensional equivalent of a centre of symmetry in three dimensions.
‡ Normally *m*; the full symbol is given so as to indicate the correct meaning of the point-group symbol.
§ The point will always correspond to the origin of the reference axes, otherwise there would be a multiplicity of origins.

It is most important to understand the reason for the change in the meaning of the positions of the point-group symbol as we progress down Table 3.1. Consider Fig. 3.4; the illustration in (a) shows a 4-fold rotation point lying on a mirror line perpendicular to the *x* axis. It is inconsistent, because the 4-fold rotation must act upon all other symmetry elements in the point group as well as on the representative points: we must proceed to (b) to achieve the necessary condition. Thus, in the symbol 4*mm*, '4' refers to the degree of rotation about a point, and the first '*m*' represents mirror lines perpendicular to both the *x* and the *y* axes: they are *equivalent*

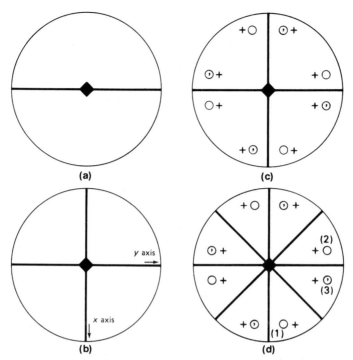

Fig. 3.4—Mirror lines in the plane point group 4mm: (a) inconsistent interaction of 4 and m, (b) consistent; m lines normal to the x and y axes are of the same *form*, (c) poles of a general form in 4m inserted, (d) stereogram (c) showing the m lines at 45° to the x and y axes, generated by the action of 4m; operations $1 \xrightarrow{4} 2$ and then $2 \xrightarrow{m} 3$ are equivalent to $1 \xrightarrow{m} 3$.

under the 4-fold rotation, or belong to the same form of mirror lines. What, then, does the second '*m*' in the symbol mean? We can answer this question by inserting a pole and operating on it with the symmetry 4m, to give illustration (c). Now we can see that two more mirror symmetry lines (another form) are implicit in the diagram, lying at 45° to the x and y axes. They can be seen in (d): they do not introduce any more symmetry-related points; they are a consequence of the interaction between 4 and m.

Such changes in the meanings of the positions of the point-group symbols from system to system can, at first, be a little confusing. However, it will prove to be a very powerful feature of the Hermann–Mauguin notation, and it carries over neatly into the realm of space groups. It cannot be stressed too strongly that a full understanding of this notation is of fundamental importance.

3.4.5 Note on symbolism

The standard symbol for reflexion symmetry in both two and three dimensions is *m*. It might be found helpful to utilize the symbol *r* for reflexion, with *R* for rotation, in two dimensions (see Appendix 2), although it is stressed that this notation is non-standard.

3.4.6 Sub-groups

It may be noted in Figs 3.1 and 3.2 that most of the point groups 'contain' point groups of lower symmetry (fewer symmetry elements). If it is possible to remove

Table 3.2—Sub-groups of the ten two-dimensional point groups

Point group	Sub-groups
1	None
2	1
3	1
4	1, 2
6	1, 2, 3
m	1
2*mm*	1, 2, *m*
3*m*	1, 3, *m*
4*mm*	1, 2, 4, *m*, 2*mm*
6*mm*	1, 2, 3, 6, *m*, 2*mm*, 3*m*

certain symmetry elements from a point group and still leave another point group, then the latter is a *sub-group* of the former. Table 3.2 lists the sub-groups of the ten two-dimensional point groups considered so far.

3.5 THREE-DIMENSIONAL POINT GROUPS

In three dimensions, the symmetry elements encountered are rotation axes, reflexion (mirror) planes and roto-inversion axes. The operations related to the first two of these symmetry elements are similar to those considered in two dimensions, except that there are now the geometrical extensions to rotation about a *line* (instead of about a point) and reflexion across a *plane* (instead of across a line). The roto-inversion axis, often called just inversion axis, has no counterpart in two dimensions.

3.5.1 Rotational symmetry in three dimensions

If an R-fold rotation about a line brings a object into an indistinguishable position for every successive rotation of $(360/R)°$, we designate that line a rotation axis of degree R. A rotational symmetry operation is often called a *proper rotation*, or an *operation of the first kind*, and is concerned with congruent parts of an object. Crystals only ever have rotation axes of degrees 1, 2, 3, 4 or 6.

3.5.2 Reflexion symmetry in three dimensions

Reflexion symmetry takes place across a plane, usually called a *mirror plane*, symbol m, such that the plane may be considered to divide the body into object and mirror-image halves.

3.5.3 Roto-inversion symmetry

An object possesses a roto-inversion axis (inversion axis) of symmetry $\bar{R}$ if it can be brought into an indistinguishable position by the combined operations of rotation through $(360/R)°$ together with an inversion through a point (see footnote § to Table 3.1) on the axis. Like mirror symmetry, roto-inversion symmetry operations cannot actually be performed on an object. Crystals have inversion axes of degrees $\bar{1}$, $\bar{2}$, $\bar{3}$, $\bar{4}$ and $\bar{6}$ only.

The mirror reflexion and the roto-inversion symmetries may be called *improper rotations*, or *operations of the second kind*, and they both involve enantiomorphous parts of an object. If the point group of a one-off three-dimensional object, be it crystal, molecule or whatever, contains $\bar{R}$ symmetry, the entire object and is enantiomorph (mirror image) are identical.

The inversion axis is a little more complex a symmetry element to contemplate than is either the reflexion plane or the rotation axis. Fig. 3.5 is a stereoview of a crystal showing a vertical $\bar{4}$ inversion axis (see Appendix 1 for stereoviewing). It is not difficult to construct a model of this crystal.

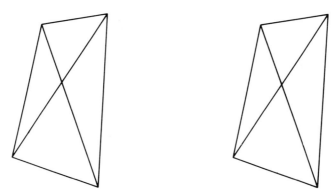

Fig. 3.5—Stereoview of an idealized tetragonal crystal with a $\bar{4}$ (vertical) axis.

A thin card is marked out according to Fig. 3.6 and then cut out along the solid lines, discarding the shaded portions. Folds are made, in the same sense, along all the dotted lines, the flaps *ADNP* and *CFLM* are glued internally, and likewise for flap *EFHJ*. Every edge of a crystal is a zone axis and thus a potential reference axis; but no two edges of this model are parallel, and there is no obvious axis to be found that way. Instead, turn the model so as to inspect it from all directions, looking for symmetry. The most obvious symmetry element is a 2-fold axis running between the mid-points of the shorter edges. Closer inspection shows that these two edges are perpendicular to each other, and that the four isosceles-triangle faces are related through a $\bar{4}$ axis (which contains 2 as a subset, or sub-group). We note also that the two mutually perpendicular planes (containing the two short edges) meet in this axis, so that one axis is so defined. We now look for others, initially hunting at right angles to the $\bar{4}$ axis. We can find two 2-fold axes, each running from the mid-point of one long edge to the mid-point of the opposite edge. These 2-fold axes are perpendicular to each other and to the $\bar{4}$ axis. So we have identified three mutually perpendicular axes (mark them on your model). As $\bar{4}$ is the 'highest order' axis, we label it z. The two diads (2-fold axes) are x and y. With axes so chosen, the point group is designated $\bar{4}2m$ (see later). We *could* have chosen the x and y axes to be lying in the m planes, while still perpendicular to $\bar{4}$ (z axis), that is, at 45° to the first choice. In that orientation, the point group is designated $\bar{4}m2$: but x and y are not then aligned along symmetry axes, and we never ignore symmetry axes when choosing reference axes for a crystal. Hence, the second option is rejected.

In Fig. 3.7, we show the operations 2 along the y axis (left to right, in the plane of the paper) and $\bar{4}$ along the z axis (normal to the plane of the paper) by means of

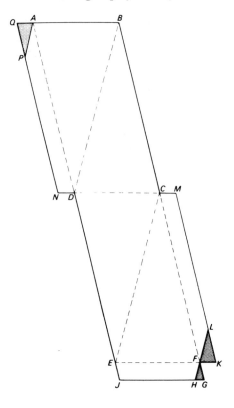

$NQ = AD = BD = BC = DE = CE = CF = KM$
$\quad = 10$ cm;

$AB = CD = EF = GJ = 5$ cm;

$AP = PQ = FL = KL = 2$ cm;

$AQ = DN = CM = FK = FG = FH = EJ = 1$ cm.

Fig. 3.6—Scheme for the construction of a tetragonal crystal with a $\bar{4}$ axis (point group $\bar{4}2m$).

the conventional stereograms (a) and (b) respectively. Fig. 3.8 shows the same stereograms, but in a revised notation used in this book: it emphasizes, first, the three-dimensional character of the object represented, by means of the + and − signs, and, secondly, the different natures of the parts represented by the points labelled (2) in (a) and (4) in (b). This feature is not so clear on the conventional stereogram, but is of importance in our use and understanding of this representation of symmetry directions. This revised notation is not essential in two dimensions.

3.5.4 Crystallographic point groups

In principle, point groups can be based on axes with any symmetry R from 1 to ∞. When other symmetry elements are combined with the first axis, their relative orientations are invariably restricted, but the total number of possible point groups is still infinite. In crystals, however, where R is restricted to 1, 2, 3, 4 and 6, the combination with other elements leads to a total of only thirty-two point groups: they are termed the crystallographic point groups. Other point groups can and do occur in flowers, molecules and man-made objects. The classificatory term *crystal class* is

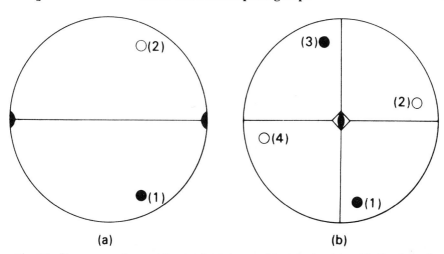

Fig. 3.7—Stereograms of general forms (a) point group 2 (axis horizontal and in the plane of the stereogram), (b) point group $\bar{4}$ (axis normal to the plane of the stereogram). In (a), the point ● is rotated through 180° to ○, $1 \to 2$. In (b), a point ● is rotated through 90° and then inverted through a point on the $\bar{4}$ axis (the origin) to ○; repetition of this combined operation, always in the same sense, generates, in all, four symmetry equivalent points $1 \to 4 \to 3 \to 2$.

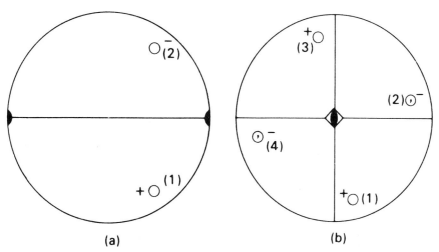

Fig. 3.8—Stereograms of Fig. 3.7 in the revised notation. The congruent parts of the object represented by ○⁺ and ○⁻ in (a) are now distinguished from the enantiomorphous parts of the object represented by ○⁺ and ⊙⁻ in (b).

not exactly synonymous with the term point group; rather, we say that crystals of a given class all belong to the same point group.

There is a simple explanation of the observation that, in crystals, R and $\bar{R}$ are restricted to 1-, 2-, 3-, 4- and 6-fold axes. Crystals can be regarded as a stack of regular parallelepipeda (solids bounded by three pairs of identical, parallelogram faces). Only solids that have one or other of the allowed symmetries can be stacked together, all in the same orientation, to fill space completely (Fig. 3.9); there are voids present for $R = 5$ and for $R \geqslant 7$.

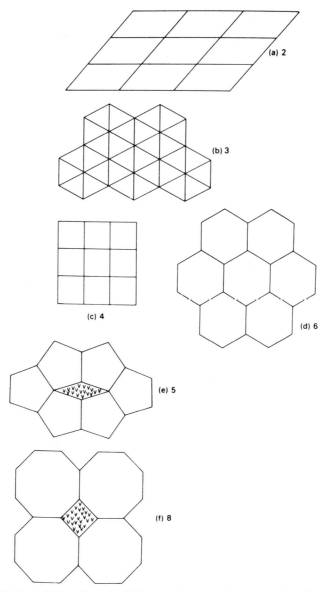

Fig. 3.9—Sections of three-dimensional figures and the rotation symmetries of their smallest structural units: (a)–(d) are space-filling arrangements, but in (e) and (f) the V-marks represent voids

In Table 3.3, some data about symmetry operations are summarized. It should be noted that $\bar{R}$ is equivalent to $R + \bar{1}$ only where R is an odd number; that operations $\bar{2}$ and m are equivalent where m is perpendicular to $\bar{2}$ (Fig. 3.10); and that the symbols for 2, 4, and $\bar{4}$ parallel to the plane of the diagram and for $\bar{2}$ are not standard, but are sometimes helpful.

In conventional stereograms, the z axis is normal to the primitive plane. Except for the triclinic system, the y axis lies in the plane of the stereogram, from left to right, and for orthogonal crystal systems the x axis runs from top to bottom in the

Table 3.3—Three-dimensional symmetry notation

Symmetry element	Written symbol	Graphic symbol	Operation on object for indistinguishable position§
Monad	1	None	360°/0° rotation about *any* line
Diad	2	● ⊥ projection ❘ ‖ projection	180° rotation about the axis
Triad	3	▲ ⊥ or inclined to projection	120° rotation about the axis
Tetrad	4	◆ ⊥ projection ■ ‖ projection	90° rotation about the axis
Hexad	6	⬟ ⊥ projection	60° rotation about the axis
Mirror plane	m	— ⊥ projection (‖ projection	Reflexion across the *m* plane
Inverse monad†	$\bar{1}$	○	360/0° rotation about the axis + inversion through a point
Inverse diad‡	$\bar{2}$	◖ ⊥ projectio ◆ ⌐ ‖ projection	180° rotation about the axis + inversion through a point
Inverse triad	$\bar{3}$	▲ ⊥ or inclined to projection	120° rotation about the axis + inversion through a point
Inverse tetrad	$\bar{4}$	◈ ⊥ projection ▯ ‖ projection	90° rotation about the axis + inversion through a point
Inverse hexad	$\bar{6}$	⬡ ⊥ projection	60° rotation about the axis + inversion through a point

† $\bar{1}$ is the operation of a centre of symmetry.
‡ $\bar{2}$ and *m* produce operations that are equivalent.
§ See footnote § to Table 3.1.

primitive plane. In the stereograms of both triclinic and monoclinic crystals, the *x* axis emerges below the plane of the paper, so that the *β* angle is, conventionally, obtuse (Fig. 3.11). For further morphological description, the reader is referred to Phillips or Tutton.†

3.5.5 Point groups and crystal systems

A broad but convenient classification of crystals may be given in terms of *crystal systems*. Crystals are allocated to the seven systems in terms of their symmetry, as indicated in Table 3.4. The *characteristic* symmetry is that which is essential to categorize a crystal; a given crystal may exhibit more symmetry than that characteristic of its system. The conventional choice of crystallographic reference axes leads to special relationships between both the intercepts *a*, *b*, *c* of the parametral plane, and the angles *α*, *β*, *γ* between them.

A good working knowledge of point groups is essential to the understanding of crystal symmetry, and it forms the basis of the ensuing study of space groups. Much of the difficulty experienced in an initial study of symmetry arises not from the nature of the symmetry operations themselves, but rather by a failure to appreciate the relative orientations of the symmetry elements given by the Hermann–Mauguin point-group symbols, and the fact that these orientations vary, necessarily, from system to system (see Section 3.4.4). A 2-fold rotation axis, for example, always produces the same type of operation. Differences arise where two or more such axes intersect, and at differing angles. Table 3.5 lists the meaning of the Hermann–Mauguin point-group symbols. It should be compared with Table 3.1, and studied carefully in conjunction with the stereogram and crystal drawings in the following section.

† See Bibliography.

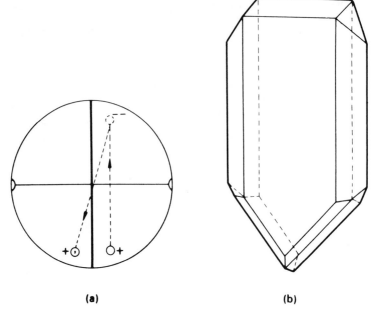

(a) **(b)**

Fig. 3.10—Stereogram to show the equivalence of the operation $\bar{2}$ and m, where the m plane is normal to the $\bar{2}$ axis: (a) stereogram, (b) crystal of point group m.

Table 3.4—Seven crystal systems

System name	Characteristic symmetry, with orientation†	Parametral plane intercepts and interaxial angles‡§
Triclinic	None (1)	$a \not\!\!\subset b \not\!\!\subset c$ $\alpha \not\!\!\subset \beta \not\!\!\subset \gamma$
Monoclinic	2 or $\bar{2}$ along y	$a \not\!\!\subset b \not\!\!\subset c$ $\alpha = \gamma = 90° \not\!\!\subset \beta$
Orthorhombic	Three mutually perpendicular 2 or $\bar{2}$ along x, y and z	$a \not\!\!\subset b \not\!\!\subset c$ $\alpha = \beta = \gamma = 90°$
Tetragonal	4 or $\bar{4}$ along z	$a = b \not\!\!\subset c$ $\alpha = \beta = \gamma = 90°$
Cubic	Four 3 along $\langle 111 \rangle$	$a = b = c$ $\alpha = \beta = \gamma = 90°$
Hexagonal	6 or $\bar{6}$ along z	$\begin{cases} a = b \not\!\!\subset c \\ \alpha = \beta = 90°; \gamma = 120° \end{cases}$
Trigonal	3 along z	
Trigonal¶	3 along $\langle 111 \rangle$	$a = b = c$ $\alpha = \beta = \gamma \not\!\!\subset 90°, < 120°$

† $\bar{2}$ is equivalent to m, where $\bar{2}$ is perpendicular to the m.
‡ We shall see in Chapter 4 that the same constraints apply to the conventional unit cells in the Bravais lattices; α, β and γ are chosen to be obtuse as far as is practicable.
§ The symbol $\not\!\!\subset$ should be read as 'not constrained by symmetry to equal'.
¶ On rhombohedral axes; here, and under Cubic, $\langle 111 \rangle$ means [111], [$\bar{1}$11], [1$\bar{1}$1] and [$\bar{1}\bar{1}$1], because $[UVW]$ and $[\bar{U}\bar{V}\bar{W}]$ are the same line.

3.6 DERIVATION OF THE CRYSTALLOGRAPHIC POINT GROUPS

Although the standard symbol m is used, conventionally, in place of $\bar{2}$, it is convenient here, as in Table 3.5, to employ $\bar{2}$, and so to consider the derivation of the crystallographic point groups in terms of the symmetry elements R and $\bar{R}$.

Table 3.5—Crystal systems and point-group symbols

System	Point groups†	Symbol meaning for each position		
		1st position	2nd position	3rd position
Triclinic	$1, \bar{1}$	All directions in crystal	—	—
Monoclinic	$2, m, \dfrac{2}{m}$	2 or $\bar{2}$ along y	—	—
Orthorhombic	$222, mm2$ mmm	2 or $\bar{2}$ along x	2 or $\bar{2}$ along y	2 or $\bar{2}$ along z
Tetragonal	$4, \bar{4}, \dfrac{4}{m}$	4 or $\bar{4}$ along z	—	—
	$422, 4mm$ $\bar{4}2m, \dfrac{4}{m}mm$	4 or $\bar{4}$ along z	2 or $\bar{2}$ along x, y	2 or $\bar{2}$ at 45° to x, y and in horizontal plane, i.e. along $\langle 110 \rangle$
Cubic	$23, m3$	2 or $\bar{2}$ along x, y, z	3 at 54°44′ to x, y, z i.e. along $\langle 111 \rangle$	—
	$432, \bar{4}3m$ $m3m$	4 or $\bar{4}$ along x, y, z	3 at 54°44′ to x, y, z i.e. along $\langle 111 \rangle$	2 or $\bar{2}$ at 45° to x, y, z i.e. along $\langle 110 \rangle$
Hexagonal	$6, \bar{6}, \dfrac{6}{m}$	6 or $\bar{6}$ along z	—	—
	$622, 6mm$ $\bar{6}m2, \dfrac{6}{m}mm$	6 or $\bar{6}$ along z	2 or $\bar{2}$ along x, y, u i.e. along $\langle \bar{1}2\bar{1}0 \rangle$	2 or $\bar{2}$ perpendicular to x, y, u, i.e. along $\langle 10\bar{1}0 \rangle$
Trigonal‡	$3, \bar{3}$	3 or $\bar{3}$ along z	—	—
	$32, 3m,$ $\bar{3}m$	3 or $\bar{3}$ along z	2 or $\bar{2}$ along x, y, u i.e. along $\langle \bar{1}2\bar{1}0 \rangle$	—

† $\dfrac{R}{m}$ always occupies a single position in a point-group symbol because, in this case, m is perpendicular to R and only one *direction* is involved. Remember also the equivalence of $\bar{2}$ and m.

‡ Given here for hexagonal axes; on rhombohedral axes, the orientations in the 1st and 2nd positions of the symbol are [111] and $\langle 1\bar{1}0 \rangle$ respectively.

3.6.1 Ten simple point groups

The first ten (simple) point groups involve a single symmetry element, and may be written as follows:

$$1, 2, 3, 4, 6, \bar{1}, \bar{2} \ (m), \bar{3}, \bar{4}, \bar{6}$$

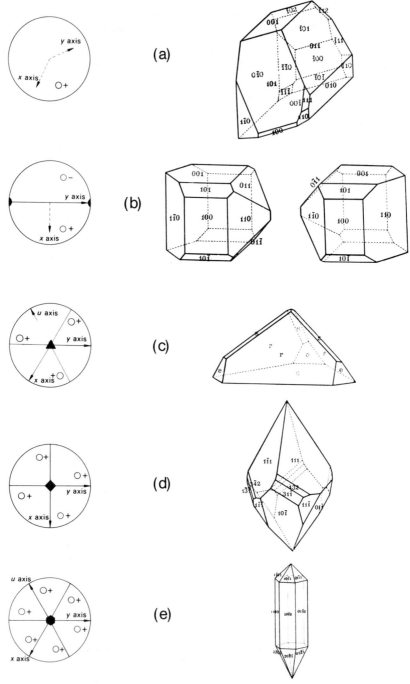

Fig. 3.11—Stereograms and crystal drawings of the ten simple point groups; the Schoenflies point-group symbols are given in parentheses after the Hermann–Mauguin symbols: (a) 1 (C_1) Calcium thiosulphate hexahydrate. (b) 2 (C_2) (+)- and (−)-Tartaric acid. (c) 3 (C_3) Sodium periodate trihydrate: c = (000$\bar{1}$), e = {02$\bar{2}$1}, r = {10$\bar{1}$1}, s = {2$\bar{1}\bar{1}$3}. (d) 4 (C_4) Wulfenite. (e) 6 (C_6) Strontium antimonyl tartrate. (f) $\bar{1}$ (C_i, S_2) Copper sulphate pentahydrate: a = {100}, b = {010}, c = {001}, o = {$\bar{1}$11}, p = {110}, p' = {1$\bar{1}$0}, p'' = {1$\bar{2}$0}, q = {011}, q' = {0$\bar{1}$1}, s = {$\bar{1}$21},

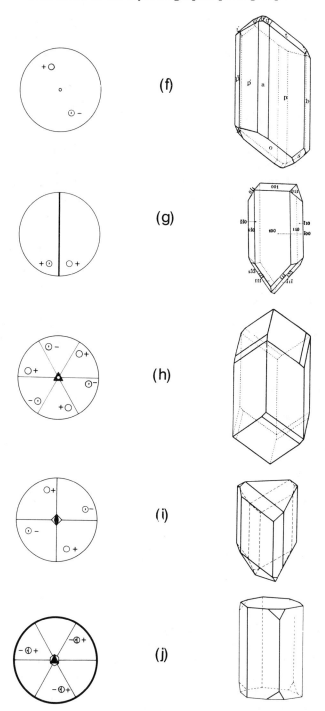

$s' = \{\bar{1}2\bar{1}\}$, $t = \{021\}$, $t' = \{0\bar{2}1\}$. (g) $\bar{2}$ (*m*) (C_s, S_1) Potassium tetrathionate. (h) $\bar{3}$ (S_6) Dioptase.
(i) $\bar{4}$ (S_4) Cahnite. (j) $\bar{6}$ (S_3) Lithium peroxide; $\bar{6}$ is equivalent to $\dfrac{3}{m}$, but the crystal is hexagonal
because $\bar{6}$ cannot operate on rhombohedral axes.

The stereograms that represent a general form in these point groups are shown together with the symmetry elements in Fig. 3.11. Further illustrations of symmetry are provided by the accompanying crystal drawings. Not all of the 32 crystallographic point groups have established representatives in the crystal kingdom; in doubtful cases, a simple morphological drawing is given. On several of the crystal drawings, the Miller indices are provided. In each of the examples, the reader should attempt to identify at least the characteristic symmetry from the disposition of the faces on the crystal.

Table 3.5 should be used to assist in the study and understanding of the orientations of all the symmetry elements shown in the crystal drawings and the stereograms. With practice, this understanding will become a sound basis of one's knowledge of crystal and molecular symmetry. Not every crystal exhibits the general form $\{hkl\}$ of its class, so that a point group deduced from the morphology alone can be

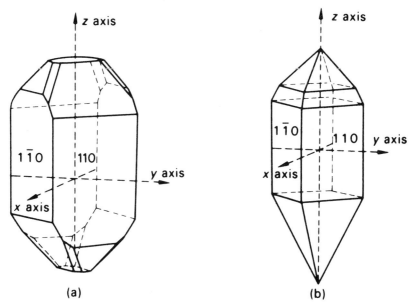

Fig. 3.12—Tetragonal crystals: (a) copper pyrites, point group $\bar{4}2m$, (b) iodosuccinimide, apparent point group $4mm$—true point group (from X-rays) 4.

misleading. Fig. 3.12 illustrates two crystals: copper pyrites shows the morphological and true point group of $\bar{4}2m$, whereas iodosuccinimide shows the morphological point group of $4mm$ but has a true point group of 4.

These ten simple point groups call for only a little further comment. Those of type $\bar{R}$ involve a change-of-hand (enantiomorphism) in the relationships to one another of the faces of a given crystal. However, only $\bar{1}$ and $\bar{3}$ are centrosymmetric. We may describe a crystal as centrosymmetric if every face is accompanied by an identical, parallel face, such that every point on one face of a pair is related to the corresponding point on the other face of that pair by inversion through the centre of symmetry. We may say that a centrosymmetric point group has $\bar{1}$ as a sub-group. The existence of an inversion axis, then, does not necessarily imply centrosymmetry: $\bar{R}$ is equivalent to $R + \bar{1}$ only where R is an odd integer; $\bar{2}$ is equivalent to m, as already explained, and $\bar{6}$ is equivalent to $\dfrac{3}{m}$; but $\bar{4}$ is unique, and cannot be represented by any other symmetry element or combination of symmetry elements.

3.6.2 Multiplicity of planes

Symmetry operations in a crystal generate (but see Section 3.4.3) crystal forms $\{hkl\}$ from the corresponding face (hkl). In a crystal of point group 4, for example, the poles o^+ in the $+x, +y, +z$ octant of Fig. 3.11(d) may be taken to represent the face (hkl). The other three symmetry-related poles then represent the faces $(\bar{k}hl)$, $(\bar{h}\bar{k}l)$ and $(k\bar{h}l)$, or collectively, the form $\{hkl\}$. This form has a multiplicity of four in this class. Note that the pairs of faces (hkl), $(\bar{h}\bar{k}l)$ and $(\bar{k}\bar{h}l)$, $(k\bar{h}l)$ are each related by 2-fold symmetry (2 is a sub-group of 4). A given form may have differing multiplicities of planes in different crystal classes, and a stereogram drawing may be used to determine the multiplicity of any plane for a given point group.

3.6.3 Combinations of symmetry elements

The combinations of R and 1, and of $\bar{R}$ and 1 are equivalent to R and $\bar{R}$ respectively, since 1 is a 360° rotation, or identity. The combinations of R and $\bar{1}$, which may be written formally as follows, produce further point groups:

$$1\bar{1}, \; 2\bar{1}, \; 3\bar{1}, \; 4\bar{1}, \; 6\bar{1}$$

We have shown already that $R + \bar{1}$ is equivalent to $\bar{R}$ where R is an odd number, so that $1\bar{1}$ and $3\bar{1}$ are, in fact, $\bar{1}$ and $\bar{3}$ respectively. The remaining three groups of this type are shown, together with appropriate crystal drawings, in Fig. 3.13. These groups are centrosymmetric, and their conventional symbols are $\dfrac{2}{m}, \dfrac{4}{m}$ and $\dfrac{6}{m}$ respectively. The operation R, where R is even, followed by the operation $\bar{1}$ is equivalent to the operation m, where R is perpendicular to m (see footnote † to Table 3.5). Symbolically, we may write (for R even)

$$\bar{1} \cdot R \equiv m \tag{3.1}$$

In Fig. 3.13(b) and (c), the convention that R is along the (vertical) z axis means that the m plane is the plane of the stereogram in these two illustrations. The notation $^-\oplus^+$ implies a horizontal m plane: $+$ to $-$ because of a change in sign in the vertical direction, and open circle to comma circle because of the enantiomorphic relationship across the mirror plane. We may note, in passing, that the multiplicity of hkl in class $\dfrac{4}{m}$ is eight: four of these planes are the same as those given in Section 3.6.2; the other four are $(hk\bar{l})$, $(\bar{k}h\bar{l})$, $(\bar{h}\bar{k}\bar{l})$ and $(k\bar{h}\bar{l})$.

It is left as an exercise to the reader to show that the combination of $\bar{R}$ with $\bar{1}$ produces no new point groups; the situation so far is summarized in Table 3.6.

3.6.4 Euler's construction

In order to derive further point groups, each R axis must be next combined with a symmetry element other than 1 or $\bar{1}$. In combining R (or $\bar{R}$) with $\bar{1}$ (or 1), there was no problem about the relative orientations of the symmetry elements, under the definition of point group (Section 3.3). If we consider the interaction of R and 2, the

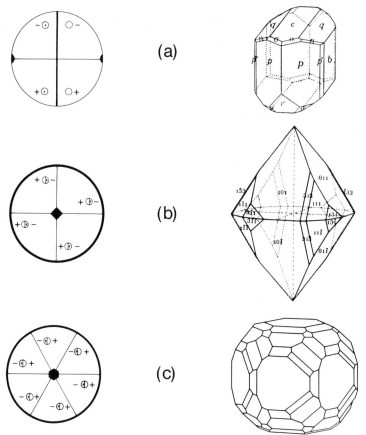

Fig. 3.13—Stereograms and crystal drawings of the three $\frac{R}{m}$ point groups: (a) $\frac{2}{m}$ (C_{2h}) Ammonium magnesium sulphate hexahydrate: b = {010}, c = {001}, n = {121}, o = {111}, o' = {$\bar{1}$11}, p = {110}, p''' = {130}, q = {011}, r' = {$\bar{2}$01}. (b) $\frac{4}{m}$ (C_{4h}) Scheelite. (c) $\frac{6}{m}$ (C_{6h}) Apatite.

Table 3.6—Thirteen of the 32 crystallographic point groups

Type	Elements	Number of new groups	Cumulative total
a	R	5	5
b	$\bar{R}$	5	10
c	$R1$	0	10
d	$\bar{R}1$	0	10
e	$R\bar{1}$	3	13
f	$\bar{R}\bar{1}$	0	13

question of their relative orientation arises immediately. It seems likely that 2 would be either parallel to R or perpendicular to it: but are these orientations correct, and are there other possibilities to consider?

It is clear that symmetry axes are zone axes. If we refer back to the section on the stereographic projection, we shall recall that a crystal can exhibit several zones, other than the trivial case of a zone of one face only (triclinic, point group 1, for example), and each of the corresponding zone axes can be identified as rotational symmetry axes of degree greater than unity. It might appear at first that a crystal could be symmetrical with respect to many different sets of intersecting axes. However, constraints exist that reduce markedly the number of possible combinations of symmetry axes that are permissible in crystals. The analysis can be carried out by a geometrical method, following Euler.

In general, two intersecting symmetry elements give rise to one or more additional mutually intersecting symmetry elements (cf. Section 3.4.4), and it is possible to show how the operation of two symmetry elements taken in succession is equivalent to the operation of another symmetry element, starting from the original position. We shall consider this proposition in terms of proper rotations, and extend it to include improper rotations.

Zone axes lie normal to their corresponding zone circles (great circles). Let OA and OB (Fig. 3.14) represent two symmetry axes intersecting in O, the centre of a spherical projection of a crystal. Let $\widehat{BAC} = \widehat{BAC'} = \alpha/2$, and let $\widehat{ABC} = \widehat{ABC'} = \beta/2$, where α and β are the angular rotations associated with the axes OA and OB respectively.

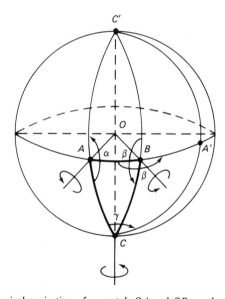

Fig. 3.14—Partial spherical projection of a crystal; OA and OB are the given symmetry axes.

Consider the motion of the point C. The anticlockwise rotation α, about OA, maps C on to C'; the anticlockwise rotation β, about OB, returns C' to its original position; thus C' may be thought of as the image of C in the plane OAB. The combination of these two rotations, $\alpha(A)$ and $\beta(B)$, leaves the point C unmoved; consequently, if there is to be a motion of any point on the sphere arising from the combination of the two rotations, the third, resultant symmetry element must pass through C', O and C.

Consider next the motion of the point A, under the same two operations. The rotation $\alpha(A)$ leaves point A unmoved: the rotation $\beta(B)$ maps A on to A', where $\widehat{A'BC} = \beta/2$; A' is, thus, the image of A in the plane OBC. In the spherical triangles ABC and $A'BC$, $\widehat{ABC} = \widehat{A'BC} = \beta/2$, $AB = A'B$, and BC is common to both triangles. The triangles are , therefore, congruent and, hence $\widehat{ACB} = \widehat{A'CB}$. Let these angles be $\gamma/2$: then the *anticlockwise* rotation γ about OC' maps A on to A'. We may write the results symbolically as

$$\beta(B)\,\alpha(A) = \gamma(C') \tag{3.2}$$

or

$$\gamma(C')\,\beta(B)\,\alpha(A) = I \tag{3.3}$$

where I represents the identity operation; (3.3) means that the successive rotations $\alpha(A)$ and $\beta(B)$, both in the same sense, followed by the rotation $\gamma(C')$ in the same sense are equivalent to the operation of identity.

We wish next to solve for the angles $\widehat{AB}$, $\widehat{BC}$ and $\widehat{CA}$ between the axes OA, OB and OC. The relevant portion of Fig. 3.14 is the spherical triangle ABC (Fig. 3.15). The solution of the triangle follows (2.46)–(2.48), to give values for a, b and c. We know that α ($=2A$), β ($=2B$) and γ ($=2C$) can take only the values $360°$, $180°$, $120°$, $90°$ and $60°$, corresponding to the rotational degrees R of 1, 2, 3, 4 and 6 respectively.

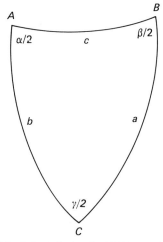

Fig. 3.15—Spherical triangle ABC from Fig. 3.14.

Formally there are 125 (5^3) solutions to (2.46)–(2.48) in our case. However, we shall apply the following constraints:

(i) $|\cos a| \leqslant 1$, $|\cos b| \leqslant 1$, $|\cos c| \leqslant 1$.

(ii) The value of $360°$ for α, β and γ is ignored, because it corresponds to the identity operation.

(iii) Since we are interested only in the number of combinations of symmetry elements, permutations will be ignored: thus, only those solutions for $\alpha = 180°$, $120°$, $90°$ and $60°$, with $\beta \leqslant \alpha$ and $\gamma \leqslant \beta$, are required.

(iv) Solutions in which one or more of a, b and c equal zero are ignored, because such results correspond to a dimensionality of less than three.

Solving the triangle subject to these conditions leads to the six non-trivial results listed in Table 3.7, and shown diagrammatically in Fig. 3.16.

The derivation of the crystallographic point groups can now be continued by considering both R and $\bar{R}$ in combinations, according to the schemes in Table

Table 3.7—Resuls following from Euler's construction

Type	α/deg	β/deg	γ/deg	a/deg	b/deg	c/deg	System
g	180	180	180	90	90	90	Orthorhombic
h	180	180	90	90	90	45	Tetragonal
i	180	180	120	90	90	60	Trigonal
j	180	180	60	90	90	30	Hexagonal
k	120	120	180	$\cos^{-1}(1/\sqrt{3})$	$\cos^{-1}(1/\sqrt{3})$	$\cos^{-1}(1/3)$	Cubic
l	90	180	120	$\cos^{-1}(\sqrt{6}/3)$	$\cos^{-1}(1/\sqrt{3})$	45	Cubic

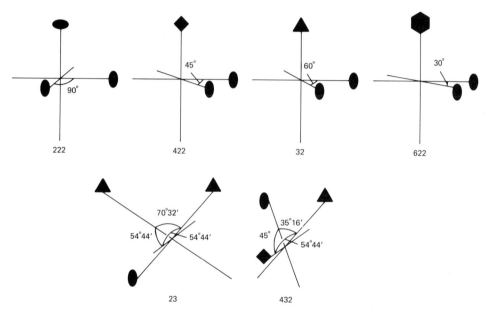

Fig. 3.16—Angles between rotation axes derived through Euler's construction.

3.5 and Fig. 3.16. Thus, for the orthorhombic system, we could write the following symbols:

$$222 \quad \bar{2}22 \quad \bar{2}\bar{2}2 \quad \bar{2}\bar{2}\bar{2}$$

$$2\bar{2}2 \quad 2\bar{2}\bar{2}$$

$$22\bar{2} \quad \bar{2}2\bar{2}$$

The symbols in the second and fourth columns are wholly inadmissible: an improper rotation, such as $\bar{2}$, combined with a proper rotation, such as 2, must lead to another improper rotation, in accordance with (3.2). The symbols in the third column are permutations, and we need choose only one of them, say, the first. Thus, we are led to the orthorhombic point groups 222 and $\bar{2}\bar{2}2$ or, in conventional notation, 222 and *mm2* (see Table 3.5). Neither of these point groups is centrosymmetric, but if we add a centre of symmetry ($\bar{1}$) to either 222 or *mm2*, we obtain *mmm* (full symbol $\frac{2}{m}\frac{2}{m}\frac{2}{m}$). Thus, for Type g in Table 3.7, we derive the three orthorhombic point groups, and stereograms and crystal drawings for them are given in Fig. 3.17.

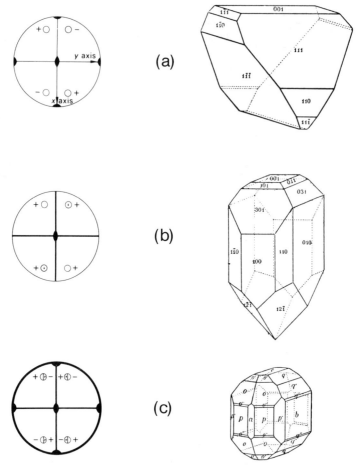

Fig. 3.17—Stereograms and crystal drawings for the orthorhombic point groups: (a) 222 (D_2)
Potassium antimonyl tartrate. (b) *mm2* (C_{2v}) Hemimorphite. (c) *mmm* (D_{2h}) Potassium sulphate:
a = {100}, b = {010}, c = {001}, o = {111}, o′ = {112}, p = {110}, p′ = {130}, q = {011}, q′ = {021},
q″ = {031}.

Continuing the derivation, Type h (Table 3.7) leads to new point groups in
the tetragonal system. Proceeding as before, we can identify the unique arrangements
as follow:

$$422, \ 4\bar{2}\bar{2}, \ \bar{4}2\bar{2}$$

or, conventionally, 422, *4mm* and $\bar{4}2m$ (Fig. 3.18). The addition of a centre of

symmetry to any one of these three point groups leads to $\dfrac{4}{m}mm$, so completing

the groups in the tetragonal system.

    Types i and j in Table 3.7 lead, by similar arguments, to the remaining trigonal
and hexagonal point groups (Figs 3.19 and 3.20), and Types k and l give the five
cubic point groups (Fig. 3.21). It is left as an exercise to the reader to derive these
twelve point groups. It may be noted that there is no third position in any of the
point-group symbols for the trigonal system (see Problem 14); also, do not confuse
32 and *3m* (trigonal) with 23 and *m3* (cubic).

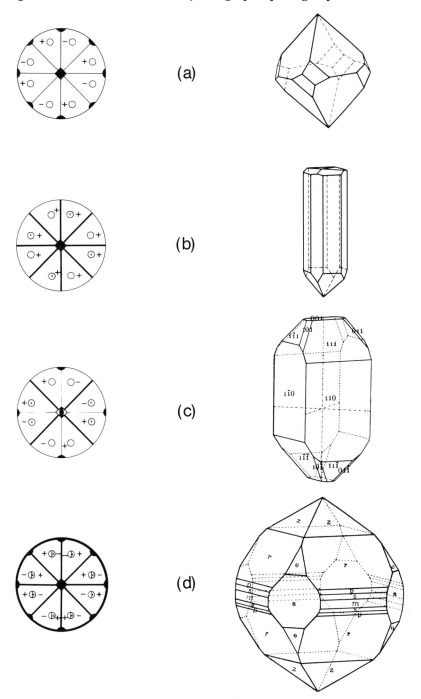

Fig. 3.18—Stereograms and crystal drawings for other tetragonal point groups: (a) 422 (D_4) Potassium hydrogen bistrichloroacetate. (b) 4mm (C_{4v}) Dibenzylthioethergold(I) chloride. (c) $\bar{4}2m$ (D_{2d}) Copper pyrites: the crystallographic axes are shown. (d) $\frac{4}{m}mm$ Anatase: a = {100}, e = {101}, m = {110}, p = {111}, r = {335}, s = {221}, z = {113}.

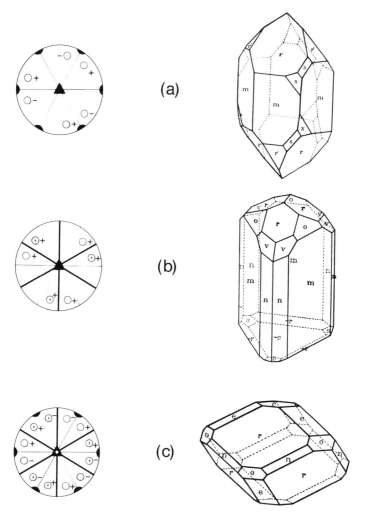

Fig. 3.19—Stereograms and crystal drawings for other trigonal point groups: (a) 32 (D_3) Quartz: m = $\{10\bar{1}0\}$, r = $\{10\bar{1}1\}$, r′ = $\{\bar{1}011\}$, s = $\{11\bar{2}1\}$, x = $\{51\bar{6}1\}$. (b) 3m (C_{3v}) Tourmaline: −c = $(000\bar{1})$, m = $\{01\bar{1}0\}$, n = $\{11\bar{2}0\}$, o = $\{02\bar{2}1\}$, r = $\{10\bar{1}1\}$, −r = $\{\bar{1}01\bar{1}\}$, s = $\{0\bar{1}12\}$, v = $\{21\bar{3}1\}$. (c) $\bar{3}m$ (D_{3d}) Calcite: c = $\{0001\}$, e = $\{01\bar{1}2\}$, n = $\{11\bar{2}0\}$, o = $\{02\bar{2}1\}$, r = $\{10\bar{1}1\}$.

The stereograms in Fig. 3.21(d),(e) show certain mirror planes as circular arcs (see Section 2.6.1); these planes are (011), (0$\bar{1}$1), (101) and ($\bar{1}$01) in a cube, and are inclined to the primitive plane in its standard orientation. It is quite easy to make a cube from a piece of white card, and to mark on it the directions of the reference axes, the Miller indices of the cube faces, and the traces and Miller indices of the inclined mirror planes.

Inclined mirror planes are not normally encountered in the stereograms of the lower symmetry groups, because it is conventional to arrange that the mirror planes are either perpendicular to or coincident with the primitive. The disposition of points related by a horizontal 4-fold axis is shown in Fig. 3.22.

No cubic point groups evolved from the simple point groups (Section 3.6.1), because the characteristic symmetry of the cubic system is the four triad axes at angles of

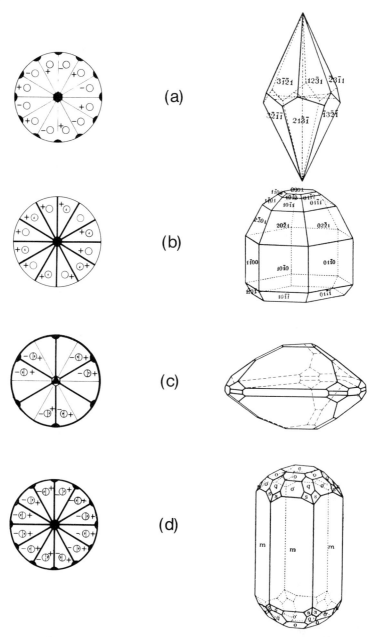

Fig. 3.20—Stereograms and crystal drawings for other hexagonal point groups: (a) 622 (D_6) Hexagonal trapezohedron $\{12\bar{3}1\}$: the crystallographic axes are shown. (b) 6mm (C_{6v}) Greenockite. (c) $\bar{6}m2$ (D_{3h}) Benitoite. (d) $\dfrac{6}{m}mm$ (D_{6h}) Beryl: c = $\{0001\}$, m = $\{10\bar{1}0\}$, o = $\{10\bar{1}1\}$, o′ = $\{20\bar{2}1\}$, q = $\{11\bar{2}1\}$, s = $\{21\bar{3}1\}$.

$\cos^{-1}(1/3)$ to one another, so that the matter of relative orientation arises immediately. The four triads incur a symmetry-equivalence of the directions of the x, y and z axes, which are two-fold symmetry axes.

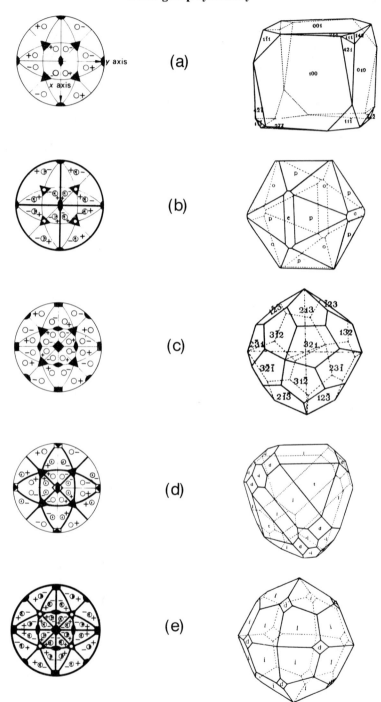

Fig. 3.21—Stereograms and crystal drawings for the cubic point groups: (a) 23 (T) Barium nitrate. (b) $m3$ (T_h) Cobaltite: c = {100}, o = {111}, p = {210}. (c) 432 (O) Pentagonal icositetrahedron {213}; the crystallographic axes are shown. Ammonium chloride has been reported to show point group 432, but it is usually classed as $m3m$. (d) $\bar{4}3m$ (T_d) Tetrahedrite: d = {110}, i = {211}, −i = {2$\bar{1}$1}, t = {111}. (e) $m3m$ Garnet: d = {110}, i = {211}.

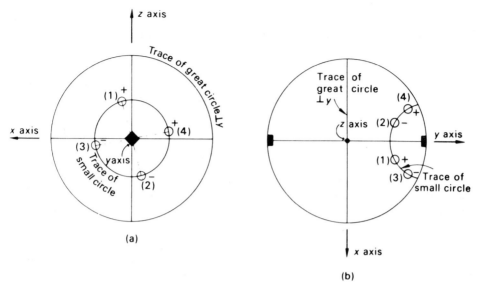

Fig. 3.22—Four-fold rotation axis in the primitive plane. (a) Four general intersections on the sphere, related by the tetrad axis along y and lying on a small circle. (b) The same four positions as poles on a stereogram in the conventional orientation (cf. Fig. 3.21(c)).

The 32 crystallographic point groups are summarized in Table 3.8; the arrows indicate the result of adding a centre of symmetry to the groups that are non-centrosymmetric. The descriptions $\bar{4}m2$ and $\bar{6}2m$ are equivalent to $\bar{4}2m$ and $\bar{6}m2$ by rotations of the x, y axes about the z axis by 45° and 30° respectively.

Table 3.8—Thirty-two crystallographic point groups†

General notation	Triclinic	Monoclinic	Trigonal	Tetragonal	Hexagonal	Cubic
R	1 ⌐	2 ⌐	3 ⌐	4 ⌐	6 ⌐	23‡ ⌐
$\bar{R}$	$\bar{1}$ ⌐	m ⌐	$\bar{3}$ ⌐	$\bar{4}$ ⌐	$\bar{6}$ ⌐	m3‡ ⌐
R + centre		$\frac{2}{m}$		$\frac{4}{m}$	$\frac{6}{m}$	
			Orthorhombic			
$R2$		222 ⌐	32 ⌐	422 ⌐	622 ⌐	432 ⌐
Rm		$mm2$	$3m$	$4mm$	$6mm$	$\bar{4}3m$
$\bar{R}m$			$\bar{3}m$	$\bar{4}2m$	$\bar{6}m2$	m3m
$R2$ + centre		mmm		$\frac{4}{m}$ mm	$\frac{6}{m}$ mm	

† The reader should consider the implications of the unfilled spaces in this table.
‡ R refers to 2 or $\bar{2}$ here, and to 4 or $\bar{4}$ below in this column.

3.6.5 Laue groups

Eleven point groups in Table 3.8, shown in bold face, are centrosymmetric. They are called Laue groups, because the X-ray diffraction Laue patterns from crystals correspond to one or other of these eleven point groups, whether or no the crystal is centrosymmetric. We shall discuss this effect in a later chapter. We may note, in passing, that there are, by analogy, six two-dimensional Laue groups: they are 2, 2mm, 4, 4mm, 6 and 6mm; they all contain the two-fold rotation point symmetry element.

3.6.6 Projected symmetry

A three-dimensional form, as illustrated by a stereogram, can be projected on to a plane, and the resulting *projected* symmetry then described by one of the plane point-groups. This projected symmetry is sometimes confused with Laue projection symmetry, and for this reason we shall defer a study of the latter until the later chapter on X-ray diffraction.

The symmetry of a projected arrangement will depend upon the plane of projection that is chosen. Table 3.9 lists the symmetries of the most important projections of the crystallographic point groups. The results are easily derived by means of stereograms. For example, consider point-group 2 (Fig. 3.11(b)). If, on the one hand, we project the poles of the general form down the direction of the z axis, that is, on to the (001) plane, the plane of projection, the resulting arrangement is similar to that of Fig. 3.1(f), that is, plane point-group m. (This is a section where the student may find the use of r instead of m in two dimensions helpful—see Appendix 2.) If, on the other hand, we project the same group on to (010), the resulting plane point-group is 2.

One may be tempted to relate the points on the (001) projection by a diad along the y axis: but the diad would rotate the points out of the plane, which is inadmissible for two-dimensional symmetry. Another consideration that may, at first, prove to be an obstacle is that the poles in a *congruent* relationship in point-group 2 project as poles in an enantiomorphic relationship in *plane* point-group m. In order to rationalize this result, consider two left hands (Fig. 3.23), one above the plane of the paper +, and that symmetry-related to it below the plane −. Let them be projected on to the plane of the paper (becoming infinitely thin); they now assume an enantiomorphic relationship, and the only possible *two-dimensional* symmetry element relating them is the mirror line, m.

In studying Table 3.9, it will be seen that the most important planes of projection may change with a change in the crystal system; Table 3.5 may be reviewed in this context.

The symbol 1 in 3m1 is used to indicate the orientation of the m line, as also in 31m. This notation is not required in the cubic point groups 23 and m3—why?

3.7 POINT GROUPS AND PHYSICAL PROPERTIES

Occasionally, it happens that the observation of a particular physical property is of use in the determination of the point group of a crystal. Generally, however, the information in this section is more of interest in terms of the effect of symmetry on physical properties.

Crystals are, generally, anisotropic in their physical properties, that is, the magnitude of the physical property varies with the direction of measurement in the crystal. Some crystals, because of their symmetry, become isotropic for certain of their properties or, perhaps, isotropic along certain zone axes. For example, cubic crystals are isotropic for the refraction of light, whereas calcite is isotropic only along [0001]. Cubic crystals are not isotropic for all physical properties; for example, they are anisotropic for elasticity measurements (see also Section 1.2).

Table 3.9—Projected point-group symmetry for the crystallographic point groups

Point group	{100}	{010}	{001}
1	1	1	1
$\bar{1}$	2	2	2
2	m	2	m
m	m	1	m
$\dfrac{2}{m}$	2mm	2	2mm
222	2mm	2mm	2mm
mm2	m	m	2mm
mmm	2mm	2mm	2mm
	{001}	{100}	{110}
4	4	m	m
$\bar{4}$	4	m	m
$\dfrac{4}{m}$	4	2mm	2mm
422	4mm	2mm	2mm
4mm	4mm	m	m
$\bar{4}2m$	4mm	2mm	m
$\dfrac{4}{m}mm$	4mm	2mm	2mm
	{0001}	{10$\bar{1}$0}	{11$\bar{2}$0}
6	6	m	m
$\bar{6}$	3	m	m
$\dfrac{6}{m}$	6	2mm	2mm
622	6mm	2mm	2mm
6mm	6mm	m	m
$\bar{6}m2$	3m1	2mm	m
$\dfrac{6}{m}mm$	6mm	2mm	2mm
3†	3	1	1
$\bar{3}$†	6	2	2
32†	31m	m	2
3m†	3m1	m	1
$\bar{3}m$†	6mm	2mm	2
	{100}	{111}	{110}
23	2mm	3	m
m3	2mm	6	2mm
432	4mm	3m	2mm
$\bar{4}3m$	4mm	3m	m
m3m	4mm	6mm	2mm

† Referred to hexagonal axes.

3.7.1 Morphology

Morphology is the study of the external forms of crystals. In principle, the representation of the goniometric measurements of a crystal on a stereogram, or other suitable projection, can reveal its true point group (Figs 3.11, 3.13, 3.17–3.21).

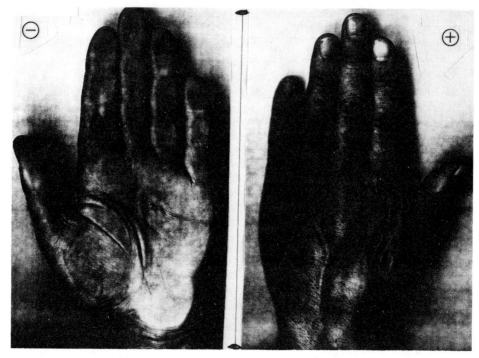

Fig. 3.23—Motif consisting of two left hands related by a two-fold axis in the plane of the diagram (point group 2). On projection, the sense of the third dimension, $\oplus$ and $\ominus$, is lost, and the diad axis degrades to a reflexion *line* (m).

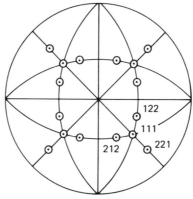

Fig. 3.24—Conventional stereogram of an alum crystal showing the octahedron {111} and the trisoctahedron {221}; cubic point groups m3 and m3m are both consistent with this diagram.

It is necessary that the general form {*hkl*} be developed on the crystal; failure in this respect may lead to a symmetry higher than true being deduced.

Crystals of alum, potassium aluminium sulphate dodecahydrate, exhibit prominent octahedral {111} development, with the trisoctahedron {221} less well developed. Both of these forms are shown on a stereogram in Fig. 3.24. From a comparison of this figure with Fig. 3.21, either of the point groups m3 or m3m could be deduced for alum. Its true point group is m3, from X-ray diffraction measurements; both the octahedron and the trisoctahedron are special forms (see Section 3.4.3) in point groups $m3$, 432 and $m3m$.

3.7.2 Optical properties

An examination of crystals with a polarizing microscope, in white light, leads to a segregation of the seven crystal systems according to the following scheme (Table 3.10).

Table 3.10—Optical classification of crystals

Optical properties	Optical class	Crystal system
Refractive index invariant with direction; no polarization colours	Isotropic	Cubic
Refractive index varies with direction; polarization colours	Anisotropic, uniaxial; isotropic along the principal symmetry axes	Tetragonal Trigonal Hexagonal
	Anisotropic, biaxial; two two general directions of single refraction	Orthorhombic Monoclinic Triclinic

Biaxial crystals may often be further segregated by their extinction effects between crossed polars. Extinction is called *straight* where the field of view is dark when a prominent edge of the crystal is parallel to the microscope eyepiece cross-wires; otherwise, extinction is called oblique. The following divisions may be noted:

Orthorhombic—straight extinction on {100}, {010}, {001} and any other plane containing the x, y or z axes or the trace thereof.

Monoclinic   —straight extinction on {100}, {001} and any other plane containing the y axis or its trace.

Triclinic   —oblique extinction on all planes.

An isotropic material under strain may become optically anisotropic. Many polymers when drawn into fibres (necked) exhibit polarization colours between crossed polars, but an X-ray examination should be used in order to confirm the presence of crystalline character.

A more detailed optical study may occasionally reveal useful symmetry information, and the reader is referred to a standard text on the subject.†

Different results are obtained from a study of the *optical activity* of crystals. If plane-polarized, monochromatic light is passed through an optically active crystal, along a direction of single refraction, the plane of polarization of the emergent light beam will be found to have been rotated, by an amount which, for a given material, depends upon both the wavelength of the light and the thickness of the crystal in the direction of the beam. A good example is sodium chlorate: this substance may be recrystallized from water, whereupon the right-handed and the left-handed crystal species can be separated by hand. Fig. 3.25 illustrates the two enantiomorphs of sodium chlorate. The optical activity is lost upon dissolution of a crystal, unless the species in solution is itself optically active. For example, sucrose is optically active both in solid and in solution.

† See Bibliography.

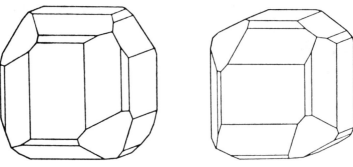

Fig. 3.25—Enantiomorphs of sodium chlorate, point group 23.

Optical activity in the solid state is confined to the eleven classes that can produce enantiomorphous crystals; they have point groups without inversion symmetry of any sort (see 3.5.3):

$$R \quad 1, 2, \quad\quad 3, \ 4, \ 6, \ 23$$

$$R2 \quad\quad 222, \ 32, \ 422, \ 622, \ 432$$

A crystal belonging to one of these eleven enantiomorphous classes is not necessarily optically active; a screw structure is a prerequisite. Thus, barium nitrate (Fig. 3.21(d)), which has the same point group as sodium chlorate, is not optically active. Some examples of optically active crystals are listed in Table 3.11

Table 3.11—Some optically active crystals

Crystal	Point group	Rotation per mm thickness in Na_D light
Quartz	32	21.7
Ethylene diamine sulphate	422	15.5
Tartaric acid	2	10.8
Sodium chlorate	23	3.1
Magnesium sulphate heptahydrate	222	2.0

3.7.3 Pyroelectric effect

It has been known for a long time that a heated tourmaline crystal attracts paper. Tourmaline belongs to class $3m$, and one end of the crystal (Fig. 3.26) is more pointed than the other end (see also Fig. 3.19(b)). The more pointed end becomes positively charged when the crystal is heated, and the other end becomes negatively charged. If the crystal is cooled, the charges are set up in the opposite sense. The development of electrical charges under thermal stress is the pyroelectric effect, and it is confined to non-centrosymmetric crystals.

Heating or cooling a crystal sets up strains in the material that, in turn, can produce a piezoelectric effect, and true pyroelectricity is not easily observed. The absence of a pyroelectric effect should not be regarded as a proof of centrosymmetry; the magnitude of the effect may be too small to be measurable.

The pyroelectric effect can be demonstrated very easily with a crystal of quartz. The crystal is suspended on a cotton thread and dipped into liquid air. On removal

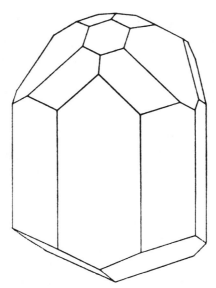

Fig. 3.26—Crystal of tourmaline, point group 3*m*. The unique axis (triad) develops + and − charges as shown when the crystal is heated.

from the liquid air, a mixture of red lead (Pb_3O_4) and sulphur is shaken on to it through a *muslin* bag. The bag charges the particles of red lead and sulphur positively and negatively respectively, and the vertical (prism) faces of the quartz crystal are coated alternatively red and yellow, according to the charges developed on the prism faces.

3.7.4 Piezoelectric effect

Certain crystals develop charges on their surfaces when subjected to mechanical stresses: this feature is the piezeoelectric effect, and simple apparatus exists for its detection†. The effect can occur only in non-centrosymmetric crystals, along the direction of a polar axis. A polar axis will have different facial development at its opposite ends of the crystal, such as tartaric acid (Fig. 3.11(b)). The piezoelectric effect is utilized in the quartz crystal oscillatory circuit. An alternating field is applied across faces of a quartz crystal plate cut either parallel to or normal to (0001), whereupon the crystal plate is set into resonant oscillation at a frequency that is inversely proportional to the thickness of the plate; the change in frequency with temperature is very small.

To summarize Section 3.7, we may say that a physical examination of a crystal, particularly by the polarizing microscope and by tests for piezoelectricity, may be helpful in allocating a crystal to its correct class.

The division of crystals among the seven crystal systems is, approximately, 50% to monoclinic, 25% to orthorhombic, 15% to triclinic, and 10% to cubic, tetragonal, trigonal and hexagonal, decreasing in that order.

† Crystal Structures Limited, Cambridge.

3.8 POINT GROUPS AND CHEMICAL SPECIES

We consider next the point groups of some chemical species. Chemical representatives of all of the crystallographic point groups are not known, and hypothetical molecules have been formulated in order to illustrate the 'missing' symmetry classes.

In addition, however, we shall encounter a number of atomic groupings that exhibit symmetry elements other than those discussed hitherto in this chapter. These additional symmetry elements give rise to non-crystallographic point groups: thus, octasulphur (S_8) has the molecular symmetry $\bar{8}2m$, but it *crystallizes* in the orthorhombic class *mmm*. Molecules described by crystallographic point groups will be discussed and illustrated in the approximate order in which they have been derived in this chapter; examples of non-crystallographic point groups are collected together at the end of this section.

The spatial natures of the various species have been brought out by stereoviews, and an attempt has been made to preserve the relative sizes of both atoms and bonds.

The information given for each species includes the chemical name, its Hermann–Mauguin point-group symbol and, in parentheses, the corresponding Schoenflies point-group symbol. It is preferable not to refer to chemical species by crystal system names, although they may share the same point-group symmetry. Thus, to say, for example, that the tetracyanonickelate(II) ion is tetragonal is poor notation, that it has tetragonal symmetry is better, but that it exhibits point group $\frac{4}{m}$ (C_{4h}) is fully descriptive of its symmetry.

3.8.1 Point groups R

Molecules of point groups of general symbol R are shown in Fig. 3.27. Any non-planar molecule of the formula-type $M(abcd)$ has point-group 1. Point groups 4 and 6 are illustrated by hypothetical species. The conformations of the A—B bonds with respect to the M—A bonds should be noted, as higher point-group symmetry would obtain if these species were planar.

Fig. 3.27—Stereoviews of molecules: in these figures, and those following, the circles are drawn to represent the approximate sizes of the atoms involved. Molecules or hypothetical molecules of point groups R: (a) Bromochlorofluoromethane, 1 (C_1). (b) Hydrogen peroxide, 2 (C_2). (c) Orthophosphoric acid, 3 (C_3). (d) $M(AB)_4$, 4 (C_4); all A–B bonds are directed below the MA_4 plane. (e) $M(AB)_6$, 6 (C_6); all A–B bonds are directed below the MA_6 plane. In this figure, and in those following, up to Fig. 3.34, the molecules are presented in the (a), (b), (c), ... order indicated by the legends, without further enumeration.

M BClF

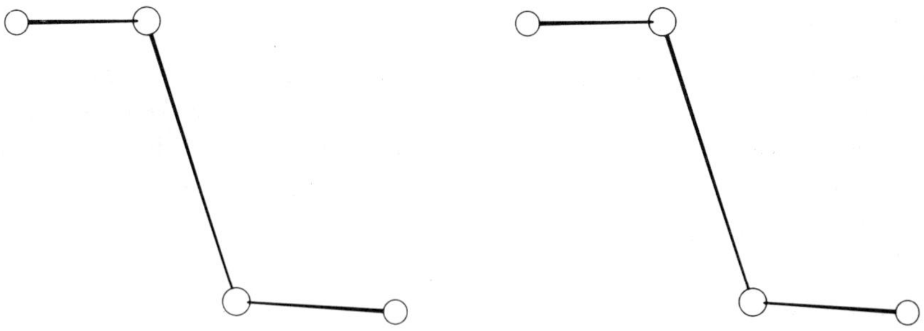

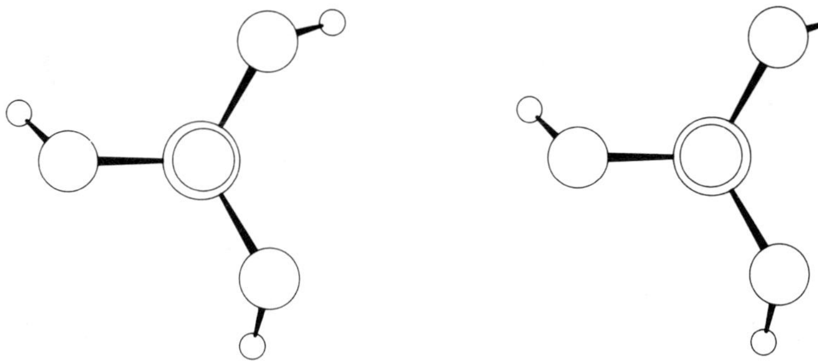

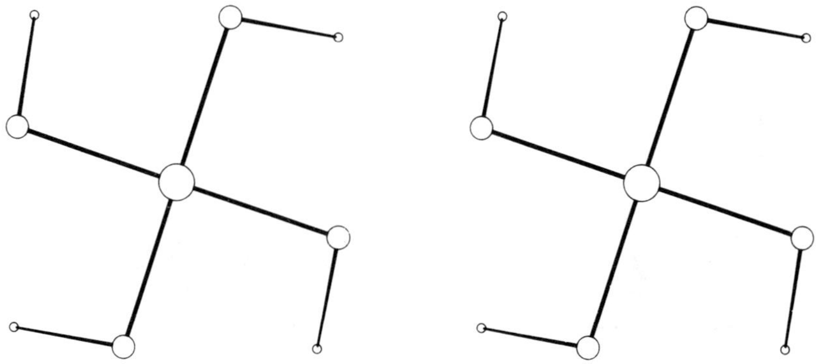

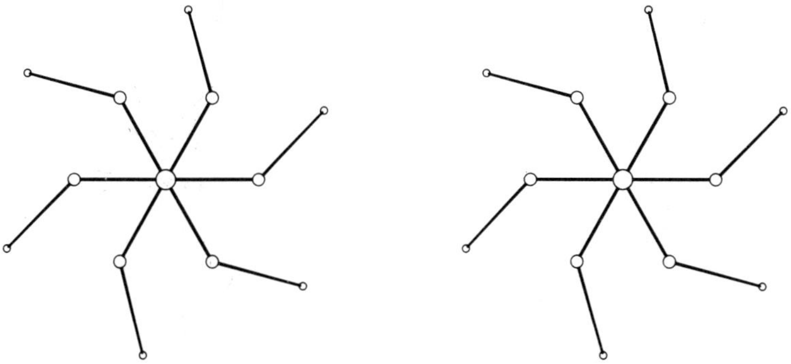

3.8.2 Point grups $\bar{R}$

All of the point groups of this type are represented by chemical species (Fig. 3.28).
As you study all of the examples in these sections, try to make sketch stereograms
to illustrate the point groups; insert the symmetry elements and the reference axes,
and note whether or no any of the atoms lie in special positions.

Fig. 3.28—Chemical species of point groups $\bar{R}$: (a) Dibenzyl, $\bar{1}$ (C_i or S_2). (b) 1,2,4-
Trichlorobenzene, $\bar{2}$ or m (C_s or S_1). (c) Hexanitronickelate(II), ion, $\bar{3}$ (S_6). (d) Dihydrogen
phosphate ion, $\bar{4}$ (S_4); the four hydrogen atom positions are occupied in a statistical manner.
(e) 1,3,5-Triazidotriazine, $\bar{6}$ (C_{3h}).

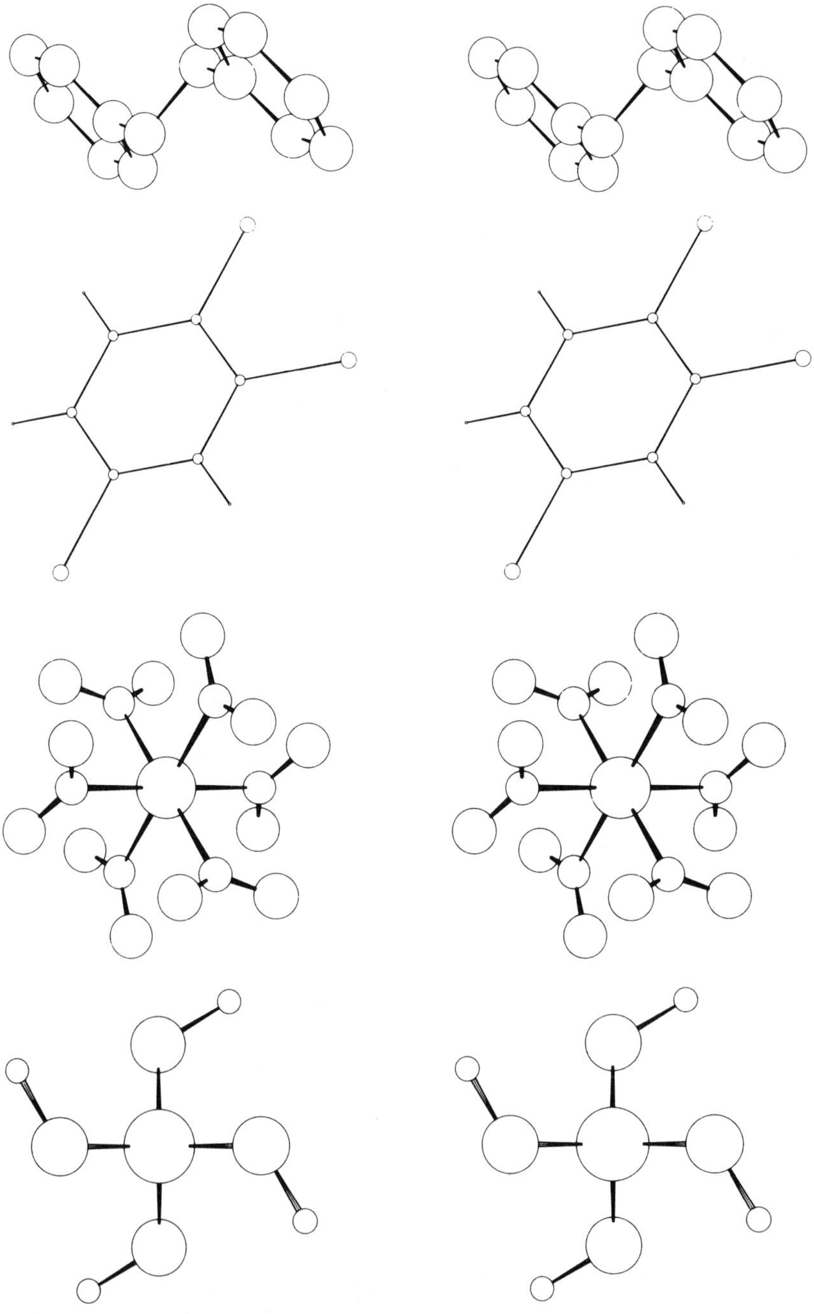

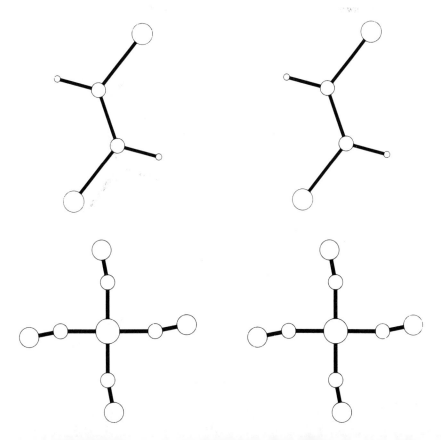

3.8.3 Point groups $\dfrac{R}{m}$

The point groups of type $\dfrac{R}{m}$ are illustrated by the species in Fig. 3.29; the molecule $M(AB)_6$ is hypothetical. All of these species are planar, that is, all of the atomic centres in each species are coplanar; the species are, of course, three-dimensional objects.

Fig. 3.29—Chemical or hypothetical species of point groups $\dfrac{R}{m}$: (a) trans-1,2-Dichloroethene, $\dfrac{2}{m}$ (C_{2h}). (b) Tetracyanonickelate(II) ion, $\dfrac{4}{m}$ (C_{4h}). (c) $M(AB)_6$, $\dfrac{6}{m}$ (C_{6h}); compare with Fig. 3.27(e).

3.8.4 Orthorhombic point groups

The orthorhombic point groups are represented by the molecules shown in Fig. 3.30. Cycloocta-1,5-diene is not a rigid molecule, and can take on conformations having symmetry other than 222 (shown here). Why not make a model of this molecule and investigate its interesting symmetries?

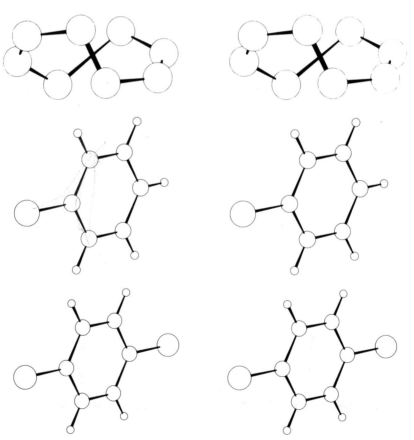

Fig. 3.30—Molecules of the orthorhombic point groups: (a) Cycloocta-1,5-diene, 222 (D_2). (b) Chlorobenzene, $mm2$ (C_{2v}). (c) 1,4-Dichlorobenzene, mmm (D_{2h}).

3.8.5 Other tetragonal point groups

Chemical representatives of these point groups are shown in Fig. 3.31. In the tetranitrodiamminocobaltate(III) ion, the unique four-fold axis is along the H_3N—Co—NH_3 direction, which requires the NH_3 groups to be in free rotation in the solid state. This interesting feature of free rotation was first noticed in a study of the propylamonium halides by Hendricks in 1928.

Fig. 3.31—Chemical species of other tetragonal point groups: (a) Tetranitrodiammino-cobaltate(III) ion, 422 (D_4); the NH_3 groups are in free rotation, and are shown as spheres. (b) Pentafluoroantimonate(III) ion, 4mm (C_{4v}). (c) Thorium tetrabromide, $\bar{4}2m$ (D_{2d}). (d) Tetrabromoaurate(III) ion, $\frac{4}{m}mm$ (D_{4h}).

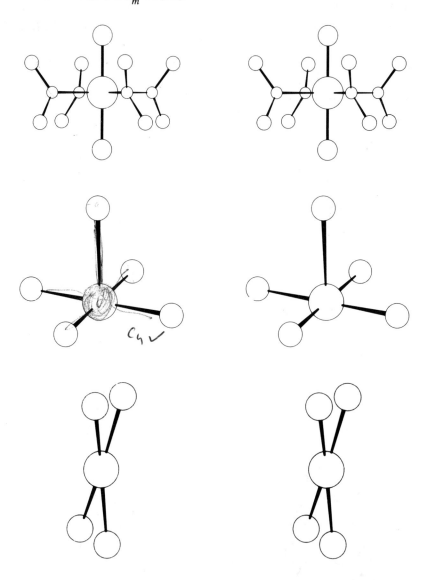

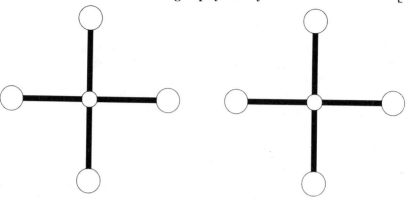

3.8.6 Other trigonal and hexagonal point groups

The remaining trigonal and hexagonal point groups are illustrated by the species in Fig. 3.32. Hypothetical molecules have been drawn for point groups 622 (compare with that for 6 in Fig. 3.27(e)) and 6mm.

Fig. 3.32—Chemical or hypothetical species of other trigonal and hexagonal point groups: (a) Dithionate ion, 32 (D_3); shown along the line of the S–S bond. (b) Trichloromethane 3m (C_{3v}). (c) Cyclohexane (chair form), $\bar{3}m$ (D_{3d}). (d) $M(AB_2)_6$, 622 (D_6). (e) MA_6B, 6mm (C_{6v}). (f) Carbonate ion, $\bar{6}m2$ (D_{3h}). (g) Benzene, $\frac{6}{m}mm$ (C_{6h}).

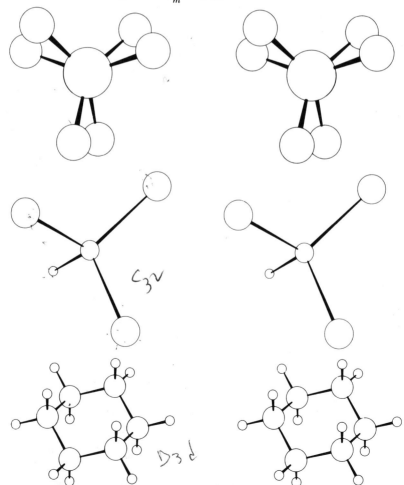

$D6$

C_6v

D_3h

C_6h

3.8.7 Cubic point groups

There are no confirmed chemical species for point groups 23 and 432, and hypothetical MA_{12} and MA_{24} molecules have been drawn (Fig. 3.33). They could be thought of in terms of the spherical projections corresponding to the stereograms of Figs 3.21(a) and (c), by regarding all poles as atoms A, and with M at the centre.

Fig. 3.33—Chemical or hypothetical species of cubic point groups: (a) MA_{12}, 23 (T). (b) Hexanitrocobaltate(III) ion, $m3$ (T_h). (c) MA_{24}, 432 (O). (d) Methane, $\bar{4}3m$ (T_d). (e) Hexachloroplatinate(IV) ion, $m3m$ (O_h).

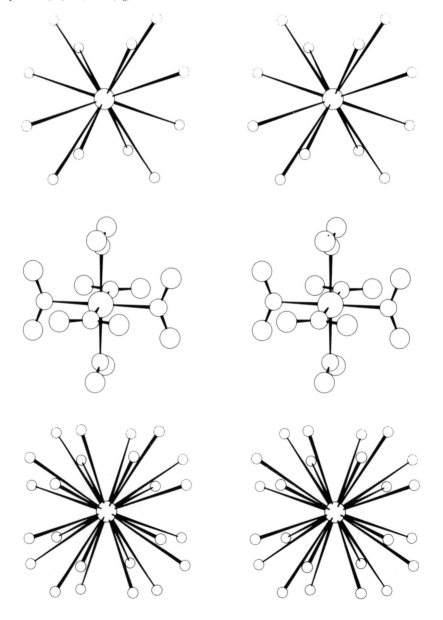

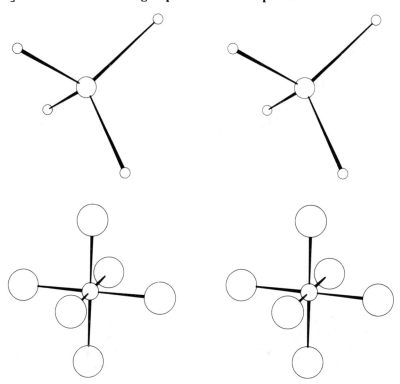

3.8.8 Non-crystallographic point groups

Fig. 3.34 illustrates chemical molecules that exhibit non-crystallographic point-group symmetry. The symmetry elements R and $\bar{R}$ have the same actions as before, but R may take values other than 1, 2, 3, 4 and 6. A linear and symmetrical molecule MA_2, such as carbon dioxide, has the point group $\bar{\infty}$, or $\dfrac{\infty}{m}$. If an MA_2 molecule is bent, such as nitrogen dioxide, its point group is $mm2$.

Fig. 3.34—Chemical species of non-crystallographic point groups: (a) Iodine monochloride, ∞m, ($C_{\infty v}$). (b) Carbon disulphide, $\dfrac{\infty}{m}$ ($D_{\infty h}$). (c) Nitrosylcyclopentadienylnickel, $5m$ (C_{5v}). (d) bisCyclopentadienyliron, $\bar{5}m$ (D_{5d}). (e) Octafluorotantalate(V) ion, $\bar{8}2m$ (D_{4d}). (f) Uranium heptafluoride, $\overline{10}m2$ (D_{5h}).

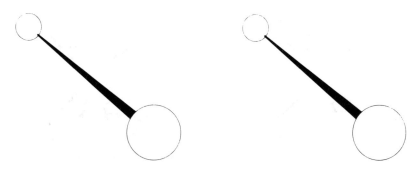

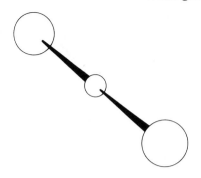
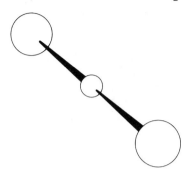

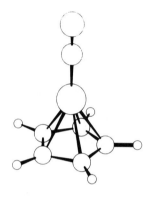
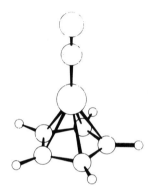

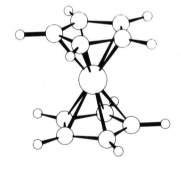
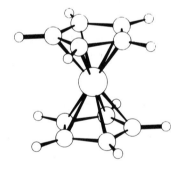

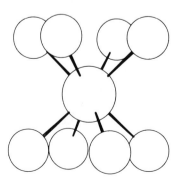
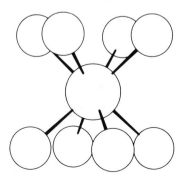

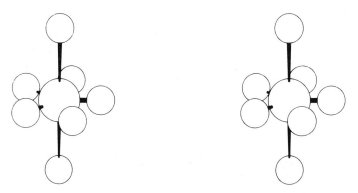

A stereogram of the bond directions in biscyclopentadienyl ruthenium is shown in Fig. 3.35; this molecule is non-centrosymmetric, but biscyclopentadienyl iron, with point group $\bar{5}m$ is centrosymmetric. Why is there no third position, other than identity, in the point-group symbol $\bar{5}m$?

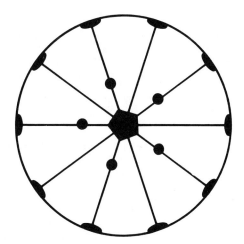

Fig. 3.35—Stereogram of the bond directions in uranium heptafluoride; the poles represent five F atoms in one plane; the other two F poles are at the centre of the stereogram; $\overline{10}$ is equivalent to $\frac{5}{m}$.

3.9 MATRIX REPRESENTATION OF SYMMETRY OPERATIONS

We conclude this chapter with a short discussion on the representation of symmetry operations and their combinations by matrix methods. This technique has a certain elegance, and is useful in establishing a relationship between point groups and space groups. A given symmetry operation can act upon any aspect of a body, such as a face on a crystal, or an atom in a molecule. In this discussion, we shall use the coordinates x y, z for a point on a three-dimensional object; for convenience, we shall represent this coordinate triplet by the vector **x**.

A symmetry operation can be written in a general way by means of the equation

$$\mathbf{R} \cdot \mathbf{x} + \mathbf{t} = \mathbf{x}' \tag{3.4}$$

where **x** and **x′** are, respectively, the coordinate triplets before and after the symmetry operation, **R** is a matrix representing the operation and **t** is a translation vector. From the definition of point group (Secton 3.3), it follows that there can be no translational operation in a point group, that is, **t** in (3.4) is zero. This condition is achieved as long as all the symmetry elements of the group pass through a point. If this were not to be the case, then two symmetry elements, such as diad axes, would be parallel; the consequence of this arrangement is illustrated by Fig. 3.36.

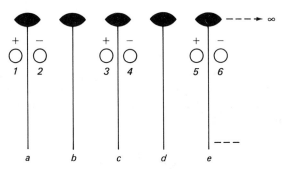

Fig. 3.36—Effect of two parallel diads *a* and *b*. Point *1* generates point *2* by rotation about axis *a*. Points *1* and *2* rotated about axis *b* produce points *3* and *4*. But *3* and *4* are now related by another diad, *c*. The effect of diad *c* on points *1* and *2* is to produce points *5* and *6*. But these points are related to *3* and *4* by diad *d* and to each other by diad *e*. Now *3* and *4*, for example, can be rotated about *e*, and so on. Clearly, this progress would lead to an infinite number of parallel, equidistant diad axes, together with the symmetry-related points, which is totally incompatible with a point group.

To take an example of (3.4), let the matrix **R**$_1$ represent the operation about a two-fold rotation axis in the monoclinic system, within the standard orientation. Then, we have

$$
\underbrace{\begin{vmatrix} \bar{1} & 0 & 0 \\ 0 & 1 & 0 \\ 0 & 0 & \bar{1} \end{vmatrix}}_{\mathbf{R}_1} \cdot \underbrace{\begin{vmatrix} x \\ y \\ z \end{vmatrix}}_{\mathbf{x}} = \underbrace{\begin{vmatrix} \bar{x} \\ y \\ \bar{z} \end{vmatrix}}_{\mathbf{x}'} \tag{3.5}
$$

Here, **R**$_1$ may be called a symmetry operator, since it represents a symmetry element with a definite orientation with respect to the reference axes.

Next, let us consider the combination of two such operators, **R**$_1$ as above, followed by **R**$_2$, a mirror reflexion across the *xz* plane. Then, we have

$$
\underbrace{\begin{vmatrix} 1 & 0 & 0 \\ 0 & \bar{1} & 0 \\ 0 & 0 & 1 \end{vmatrix}}_{\mathbf{R}_2} \cdot \underbrace{\begin{vmatrix} \bar{x} \\ y \\ \bar{z} \end{vmatrix}}_{\mathbf{x}'} = \underbrace{\begin{vmatrix} \bar{x} \\ \bar{y} \\ \bar{z} \end{vmatrix}}_{\mathbf{x}''} \tag{3.6}
$$

Consider (3.5) and (3.6) in the light of Fig. 3.11(b) and (g) respectively, and the combination of (3.5) and (3.6) in the light of Fig. 3.13(a). Schematically, we have

$$\mathbf{x} \xrightarrow[\;y\;]{\text{2 along}} \mathbf{x}' \xrightarrow[\text{to } y]{m \text{ normal}} \mathbf{x}''$$

$$\bar{1} \text{ at intersection of 2 and } m$$

in other words, we have illustrated the relationship

$$m \cdot 2 = \bar{1} \tag{3.7}$$

Another way of reaching the same result is first to combine the two symmetry operators $\mathbf{R}_1$ and $\mathbf{R}_2$ to produce $\mathbf{R}_3$:

$$
\begin{vmatrix} 1 & 0 & 0 \\ 0 & \bar{1} & 0 \\ 0 & 0 & 1 \end{vmatrix} \cdot
\begin{vmatrix} \bar{1} & 0 & 0 \\ 0 & 1 & 0 \\ 0 & 0 & \bar{1} \end{vmatrix} =
\begin{vmatrix} \bar{1} & 0 & 0 \\ 0 & \bar{1} & 0 \\ 0 & 0 & \bar{1} \end{vmatrix} \tag{3.8}
$$

$$\mathbf{R}_2 \qquad \cdot \qquad \mathbf{R}_1 \qquad = \qquad \mathbf{R}_3 \tag{3.9}$$

Equation (3.9) is an alternative presentation of the results of the combination of symmetry operations (3.2), and (3.8) is its justification in matrix manipulations.

It should be noted that (3.9) corresponds to carrying out first the operation $\mathbf{R}_1$ (on $\mathbf{x}$) and *then* the operation $\mathbf{R}_2$. In symmetry operations with $R \leqslant 2$ this rule is not important, but it is good practice nevertheless always to multiply the matrices in the standard order.

In order to illustrate this rule, consider the tetragonal point group $4mm$. If we are given the orientation of 4 and m, we can deduce the nature and orientation of the resultant m by the effect of $4m$ on a given point x, y, z, as well as obtaining the coordinates of the new point, as follows:

$$m \perp x \qquad 4 \text{ along } z$$

$$
\begin{vmatrix} \bar{1} & 0 & 0 \\ 0 & 1 & 0 \\ 0 & 0 & 1 \end{vmatrix} \cdot
\begin{vmatrix} 0 & \bar{1} & 0 \\ 1 & 0 & 0 \\ 0 & 0 & 1 \end{vmatrix} =
\begin{vmatrix} 0 & 1 & 0 \\ 1 & 0 & 0 \\ 0 & 0 & 1 \end{vmatrix} \tag{3.10}
$$

$$\mathbf{R}_2 \qquad\qquad \mathbf{R}_1 \qquad\qquad \mathbf{R}_3$$

So the point x, y, z operated upon first by $\mathbf{R}_1$ and then by $\mathbf{R}_2$ becomes y, x, z, that is, $\mathbf{R}_3$ represents an m plane normal to $[1\bar{1}0]$—refer to Fig. 3.18(b). If we had multiplied in the reverse order

$$
\begin{vmatrix} 0 & \bar{1} & 0 \\ 1 & 0 & 0 \\ 0 & 0 & 1 \end{vmatrix} \cdot
\begin{vmatrix} \bar{1} & 0 & 0 \\ 0 & 1 & 0 \\ 0 & 0 & 1 \end{vmatrix} =
\begin{vmatrix} 0 & \bar{1} & 0 \\ \bar{1} & 0 & 0 \\ 0 & 0 & 1 \end{vmatrix} \tag{3.11}
$$

$$\mathbf{R}_1 \qquad\qquad \mathbf{R}_2 \qquad\qquad \mathbf{R}_4$$

so that x, y, $z \xrightarrow{\mathbf{R}_1 \cdot \mathbf{R}_2} \bar{y}$, $\bar{x}$, z. The point $\bar{y}$, $\bar{x}$, z is related to x, y, z under the symmetry $4m$, and the m plane represented by $\mathbf{R}_4$ is normal to $[110]$, a direction in the same form as $[1\bar{1}0]$ under $4m$; therefore, $\bar{y}$, $\bar{x}$, z is one of the eight points related by the symmetry $4m$. However, if we are expecting the result $\mathbf{R}_3$ and obtain instead $\mathbf{R}_4$ it may be confusing and, in treating some properties, significantly different.

Further applications of matrix methods are given in Appendix 5, and we shall consider them again in the chapter on space groups.

PROBLEMS 3

1. (a) Draw stereograms to show a general form in each of the plane point groups 3, 3m and 4mm. Insert the symbols for the symmetry elements and indicate the directions of the conventional reference axes.

   (b) Why is m symmetry not possible in the oblique system, for example, in a parallelogram?

2. Draw stereograms to show both the symmetry elements and the general form for each of the point groups mm2, $\bar{4}2m$ and 23. What are the sub-groups for each of these point groups? What point groups are obtained by adding a centre of symmetry to each of the point groups in turn?

3. Write down the full meaning conveyed by each of the three-dimensional point-group symbols m, 422, $\bar{6}m2$ and $\bar{4}3m$.

4. Use equations (2.46)–(2.48) to derive the angles at which the symmetry axes intersect in point groups 522 and $\bar{8}2m$ ($\bar{8}2\bar{2}$). Hence, give a full meaning to each of these point-group symbols.

5. Show with the aid of stereograms that $42\bar{2} \equiv \bar{4}22 \equiv \dfrac{4}{m} mm$.

6. Take the cover of a *plain* matchbox (Fig. P3.1(a)). Write down its point group. Squash the cover diagonally (Fig. P3.1(b)). What is its point group now?

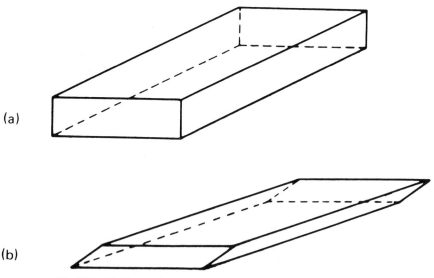

Fig. P3.1—Plain matchbox cover: (a) normal, (b) squashed diagonally.

7. What are the multiplicities of planes in the forms {010}, {110} and {123} in each of the point groups mm2, 4mm and 432? Indicate both special (s) and general (g) forms in the results.

8. What is the projected point-group symmetry for each of the following situations, and what are the Laue groups for (a) to (l)?

	Point group	Orientation			Point group	Orientation
(a)	$\bar{1}$	{100}		(g)	3	{11$\bar{2}$0}
(b)	m	{010}		(h)	$3m$	{10$\bar{1}$0}
(c)	mmm	{120}		(i)	$\bar{6}$	{0001}
(d)	$\bar{4}$	{110}		(j)	$\bar{6}m2$	{0001}
(e)	422	{110}		(k)	23	{111}
(f)	$\bar{4}2m$	{001}		(l)	432	{110}

9. Determine as much as possible about the point-group symbol in each of the following situations:

(a) Laue group $\bar{3}$; optically active.
(b) Optically biaxial; point-group 2 in projection on to (010).
(c) Projected symmetry $3m$ on (111) and m on (110).
(d) Optically active and uniaxial; m symmetry when projected on to (10$\bar{1}$0).

10. Write down the point-group symbol for each of the following derivatives of benzene.

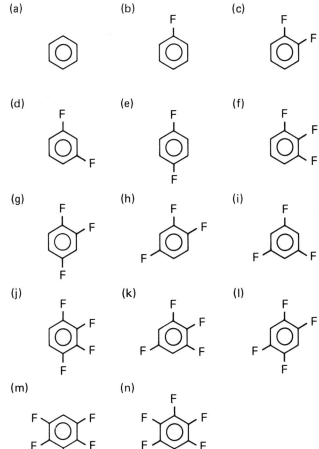

11. What are the point groups of the following molecules?

$$CH_4, CH_3Cl, CH_2Cl_2, CH_2ClBr, CHFClBr$$

12. What is the point-group symmetry of the two-dimensional figure that can be obtained by packing together (a) two (b) four irregular but identical quadrilaterals?

13. In iodine heptafluoride, IF_7, five fluorine atoms from the apices of a regular pentagon, with the iodine atom at its centre. The two remaining fluorine atoms lie on the line F—I—F normal to the plane of the pentagon. Draw a stereogram to show the directions of the I—F bonds. On another copy of the stereogram, insert the symmetry elements present in the molecule. (All I—F bonds are equal in length.) What is the point group of IF_7? What is the multiplicity of a general form in this point group, and what comment can you make about the locations of the fluorine atoms?

14. Following Section 3.9, set up matrices for the following symmetry operations: $\bar{4}$ along the z axis, and m normal to the x axis. Hence, determine the coordinates of a point obtained by operating on another point x, y, z by $\bar{4}$ and then on the resulting point by m. What are the nature and the orientation of the symmetry element obtained by this combination of $\bar{4}$ and m?

15. Set up matrices for a three-fold rotation about the z axis and for a mirror reflexion across a plane normal to the y axis for a trigonal crystal referred to hexagonal axes. Consider the combination of 3 followed by m, in the orientation given, and explain how there is no non-trivial third position in the point-group symbols for crystals in the trigonal system.

4

Lattices

You cannot make a man by standing
a sheep on its hind legs. But by
standing a flock of sheep in that
position you can make a crowd of men.

Max Beerbohm, 1911 *Zuleika Dobson*

4.1 INTRODUCTION

The geometrical basis of a crystal structure is its lattice, so that any structure may be described in terms of its three-dimensional lattice or Bravais lattice. In order to begin our appreciation of lattices we consider Fig. 4.1(a), which is one row of bricks from Fig. 1.5. We may identify a fixed point on each brick, and so obtain a row of points representative of the row of bricks, as in Fig. 4.1(b). The row of points is characterized by a constant spacing in the direction of the row.

We may next take a number of such rows of points and arrange them at a constant second spacing, so as to give a two-dimensional net of points. If we set off successive rows by one-half of the horizontal spacing, we obtain the net of points in Fig. 4.1(c). If we now replace each dot by a brick in the same orientation as in (a), we generate the *pattern* of the wall shown in Fig. 1. 5. This discussion may be extended geometrically by taking a number of nets and arranging them regularly with respect to a third dimension, so as to build up the three-dimensional lattice in Fig. 4.2.

We have looked briefly at one-, two- and three-dimensional lattices, and we may now give a formal definition of a lattice as *a regular arrangement of points, of infinite extent, such that each point is in the same vector environment as every other point.* There are several aspects of this definition that are worthy of consideration at this stage.

First, we are concerned with *regular* arrangements of points, whether in one, two- or three-dimensional space, because the essence of the crystalline nature of most solids is related to atomic regularity. Secondly, a lattice is, conceptually, of *infinite extent*. Of course, the bricks of a wall or, indeed, the basic structural units of a finite crystal are not infinite in number. However, lattice theory can be applied in both cases because in a crystal of experimental size (say, 0.1 to 0.2 mm side) examined by an X-ray diffraction probe, and in a brick wall of adequate size (say, 50 m side) examined

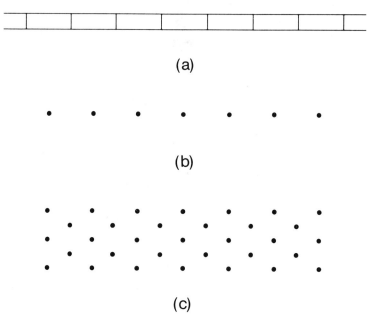

Fig. 4.1—Simulation of lattices: (a) row of bricks from Fig. 1.5, (b) row of representative points, one for each brick, taken at the same fixed location on each brick, (c) identical rows of points stacked regularly with respect to a second dimension; replacing the points with bricks, all in the same orientation as in (a), leads to Fig. 1.5.

Fig. 4.2—Three-dimensional lattice (stereoview), obtained by stacking identical nets of points regularly with respect to a third dimension.

optically by eye (visual probe), there are sufficient units to 'fool' the probe into a feeling of infinite range, much more so in the case of the crystal than of the wall.

Finally, the identical *vector environment* of the points refers to the building up of the wall from the bricks placed at the lattice points, in a fixed orientation, so that we can achieve Fig. 1.5 from the pattern of Fig. 4.1(c) and the appropriate number of bricks. This feature of the vector environment will arise again when we consider the numbers of lattices that are possible in one-, two- and three-dimensional space.

4.2 ONE-DIMENSIONAL LATTICE

In considering the number of lattices possible in any dimension, we are concerned with the number of different *arrangements* of points. Thus, differences in the amount of the (regular) spacing of the lattice points does not lead to different arrangements. Conequently, there is only one one-dimensional lattice (Fig. 4.1(b)), and the symmetry at each lattice point is that of reflexion across the point. Here, we may recall Section 3.5, and note that reflexion across a point is a geometrical contraction with respect to two dimensions; for convenience at this stage, we give the summary in Table 4.1.

Table 4.1—Geometrical extension of symmetry in one, two and three dimensions

Operation	Dimension		
	One	Two	Three
Reflexion	Across point	Across line	Across plane
Rotation	—	About point	About line
Inversion	—	—	Through point

4.3 TWO-DIMENSIONAL LATTICES

A study of the two-dimensional lattices forms a useful basis for the main topic of this chapter, the three-dimensional (Bravais) lattices.

A two-dimensional lattice, or net, has already been described (Fig. 4.1(c)) in terms of the stacking of identical rows of points in a regular manner. The most general net is shown in Fig. 4.3. It is convenient to represent the net by an appropriate portion of it, such as the unit cell I outlined by the two vectors **a** and **b**. Evidently, the net may be reconstructed by the stacking of identical unit cells, side-by-side, all in the same orientation. The unit cell shown is primitive, symbol *p*, which means that one lattice point is ascciated associated with the unit cell area. This fact may be appreciated either by taking note of the fact that each lattice point is shared equally by four adjacent unit cells, or by imagining the unit cell framework translated by a small diagonal distance, whereupon there is clearly one point per unit cell area (the thin lines in Fig. 4.3). It is helpful to remember that the term *lattice* refers to the arrangement of points; the unit cell framework is a convenience, inserted in order to permit a more ready appreciation of the geometry of the lattice than do the points alone (cf. Figs 4.1(c) and 4.4(b)).

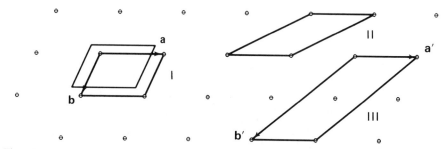

Fig. 4.3—Oblique two-dimensional lattice; I (conventional), II and III are three of an infinite number of possible unit cells. A diagonal displacement of the framework of cell I shows that there is one lattice point per unit area in a *p* cell. Similarly for II *p*, but III *c* contains two lattice points per unit area.

Each lattice point is a position of point-group symmetry. In the case of a net, the symmetry at each point will be that of one of the two-dimensional point groups. It is desirable not to refer to the point group of a unit cell. The unit cell is a representative portion of an infinite array of points and is, therefore, not amenable to description by a point-group symbol alone.

In the lattice, we encounter translation symmetry, by virtue of the points repeated identically at intervals of a and b along the x and the y axes respectively. This translation is implied by the symbol p designating the type of unit cell chosen to specify the lattice.

4.3.1 Choice of unit cell

The choice of a unit cell is, to some extent, arbitrary; Fig. 4.3 shows unit cells II and III, in addition to I, both of which represent the lattice adequately. The universal crystallographic convention is to choose the smallest repeat unit of the lattice, provided that the vectors delineating that unit are parallel to the important symmetry directions in the lattice. Thus, the conventional unit cell is not always the smallest repeat unit; sometimes the preferred unit cell is centred, symbol c.

4.3.2 Nets in the oblique system

The symmetry at any lattice point indicates an appropriate two-dimensional system for that lattice, following Table 3.1. In the conventional unit cell I of Fig. 4.3, we have $a \, ⊄ \, b$ and $\gamma \, ⊄ \, 90°$ or $120°$ ($\gamma = \widehat{\mathbf{ab}}$); the values of $90°$ and $120°$ would lead to a higher lattice symmetry. Unit cell II is another p unit cell; it has the same area as does I but the angle γ is more obtuse, which is generally less convenient. The unit cell III is centred, c; it has twice the area of I or II, and would not normally be chosen.

The centred cell III does not represent a new lattice in the oblique system because the transformation, with respect to its origin,

$$\mathbf{a} = \mathbf{a}' \tag{4.1}$$

$$\mathbf{b} = \mathbf{a}'/2 + \mathbf{b}'/2 \tag{4.2}$$

leads again to the unit cell I. This result is not surprising: the lattice and it symmetry are invariant under choice of unit cell.

4.3.3 Nets in the rectangular system

When any one of the quantities a, b or γ is specialized in a non-trivial manner, the symmetry at each lattice point will be higher than 2. (The condition $a' = 2a$ is an example of a trivial specialization.) Consider the net in Fig. 4.4: the symmetry at each point is $2mm$, and the net belongs to the rectangular system. The conventional unit cell is I, p; cell II is also p and, therefore, of the same area as I. However, cell II, taken in isolation as the representative unit of the lattice, does not obviously indicate the symmetry $2mm$ at each point. A similar argument applies to cell III. Hence, in accordance with the convention for the choice of unit cell, we must select cell I.

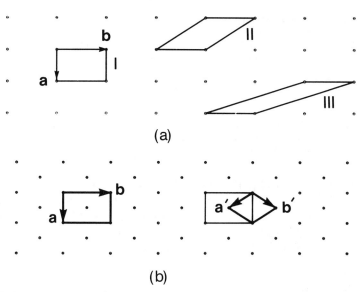

(a)

(b)

Fig. 4.4—Rectangular two-dimensional lattices: (a) conventional p cell I—$\mathbf{a}$ and $\mathbf{b}$ are parallel to the m symmetry lines in the lattice; p cells II and III have the same area as cell I but neither is the conventional unit cell; (b) cell $\mathbf{a}$, $\mathbf{b}$ is the conventional unit cell; $\mathbf{a}'$, $\mathbf{b}'$ is a possible p cell, but it is not the conventional choice.

In the oblique system, we looked at a centred cell (Fig. 4.3, III) and concluded that it did not constitute a new arrangement of points. We had $a' \mathbb{C} b'$ and $\gamma' \mathbb{C} 90°$ or $120°$, the same type of specification as for cell I.

In the case of the rectangular c cell (Fig. 4.4(b)), the primitive unit cell $\mathbf{a}'$, $\mathbf{b}'$, of smaller area, does not reveal the lattice symmetry clearly. That the rectangular symmetry is present can be inferred from the transformation equations

$$\mathbf{a}' = \mathbf{a}/2 - \mathbf{b}/2 \qquad (4.3)$$

$$\mathbf{b}' = \mathbf{a}/2 + \mathbf{b}/2 \qquad (4.4)$$

whence

$$a' = b' = (a^2/4 + b^2/4)^{1/2} \qquad (4.5)$$

Thus, a parallelogram unit cell in which its sides are equal has the symmetry $2mm$ at each lattice point (rectangular system); the value of γ' depends on the ratio a/b. This result is important because one might, on first sight, identify such a parallelogram with an oblique net.

4.3.4 Nets in the square and hexagonal systems

Further specialization of a, b and γ leads to the symmetries $4mm$ (square) and $6mm$ (hexagonal) at lattice points. The results for the five two-dimensional lattices are summarized in Table 4.2, and further discussion arises through Problem 4.1. Table 4.2 may be considered as an enhancement of the table of two-dimensional systems (Table 3.1).

Table 4.2—Two-dimensional lattices

System	Unit cell symbol	Symmetry at each lattice point†	Unit cell edges and angles
Oblique	p	2	$a \not= b; \gamma \not= 90°, 120°$
Rectangular	p, c	$2mm$	$a \not= b; \gamma = 90°$
Square	p	$4mm$	$a = b; \gamma = 90°$
Hexagonal	p	$6mm$	$a = b; \gamma = 120°$

† All point groups here have 2 as a sub-group.

The 'honeycomb' array of points (Fig. 4.5(a)) is not a lattice. The vector environments of all points are not identical, as is shown by the unequal vectors $\mathbf{p}_1$ and $\mathbf{p}_2$. A true net may be formed by centring the honeycomb (Fig. 4.5(b)), but each centred hexagon now encompasses three primitive (triply primitive) hexagonal unit cells of the type specified in Table 4.2.

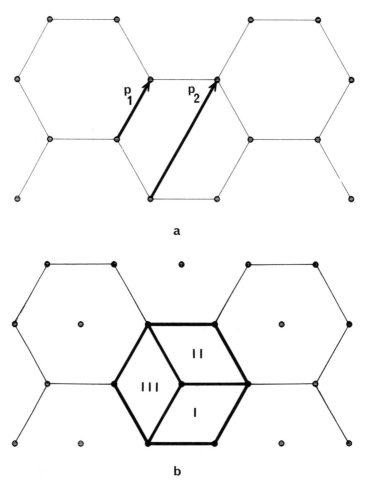

Fig. 4.5—Honeycomb arrangement of points: (a) this is not a lattice; compare $\mathbf{p}_1$ and $\mathbf{p}_2$, (b) true lattice, but each hexagon encloses three conventional hexagonal unit cells.

4.3.5 Centring

In addition to the primitive unit cells, we have considered centred unit cells in which there is a lattice point at the geometrical centre of an otherwise primitive unit cell. Only the centre of the unit cell forms a suitable site in the context of a lattice. In Fig. 4.6, we can see a regular array of points, capable of indefinite extension, but without identical vector environment for each point. Only where the additional lattice point occupies a true centring position in a unit cell can that represent a lattice. Fig. 4.6 can be described by a rectangular *p* unit cell with *two* 'points', such as *O* and *P*, associated with each lattice point in the same orientation in each unit cell. Here, we have the beginning of structure, which we shall consider more fully later on.

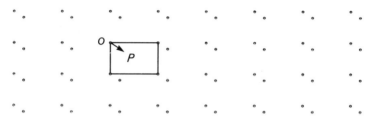

Fig. 4.6—Regular arrangement of points, of unlimited extent, but not a lattice: vector **OP** placed with its tail at *P* does not terminate on another lattice point.

4.4 THREE-DIMENSIONAL LATTICES

In our introduction (Section 4.1), we considered the general formation of a three-dimensional lattice by the stacking of nets **a**, **b** in a regular manner with respect to the third dimension **c**. There are fourteen such regular arrangements of points in three-dimensional space, as pointed out by Bravais in 1848. Frankenheim, in 1842, had described fifteen arrangements, but two of them were shown by Bravais to be identical. For this 'sin', Frankenheim's name is generally omitted from a discussion of lattices, and we speak of the fourteen Bravais lattices.

The three-dimensional lattices are distributed, unequally, among the seven crystal systems according to the lattice symmetry. Each lattice is represented by a unit cell, itself specified by three non-coplanar vectors **a**, **b** and **c**, which are chosen so that they form a parallelepipedon of smallest volume subject to the condition that the unit cell edges are parallel to the important symmetry directions in the lattice (compare Section 4.3.1). In this way, we select a convenient cell, not always primitive, that displays clearly the symmetry of the lattice.

It may be noted that the common terms such body-centred lattice and face-centred lattice are misuses of the word *lattice*. A lattice itself cannot be centred in any way; the centring terminology implies a choice of unit cell.

4.4.1 Triclinic lattice

Fig. 4.7 represents a triclinic lattice. The unit cell is characterized by the conditions $a \neq b \neq c$ and $\alpha \neq \beta \neq \gamma \neq 90°, 120°$. It may be noted that similar conditions were given for the intercepts of the parametral plane and for the interaxial angles given in Table 3.4, arising from purely morphological considerations.

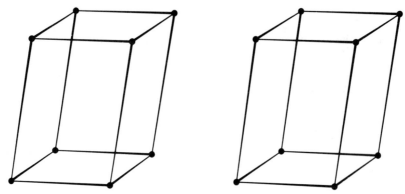

Fig. 4.7—Stereoview of a *P* unit cell in the triclinic lattice. In this illustration, and all those of unit cells up to and including Fig. 4.27, the orientation (standard) of the unit cell is that with the origin in the bottom, rear left corner so that $+a$ is directed towards the reader, $+b$ to the right and $+c$ upwards. The ratios $a:b:c$ and the values of α, β and γ are given in Table 4.4, and following Fig. 2.7.

There is only one triclinic lattice. Its conventional unit cell is primitive, symbol *P* (upper case letters for three dimensions), and the symmetry at each lattice point is $\bar{1}$. Any valid centred triclinic unit cell, that is, one which represents a true triclinic lattice, can be reduced to a *P* cell which is the more conventional. Any *P* unit cell contains one lattice point per unit volume (compare Section 4.3).

When the vectors **a**, **b** and **c** are specialized in a non-trivial manner, the lattice has symmetry higher than $\bar{1}$ at each point. We shall investigate apparently different centred unit cells in some of the crystal systems and show, for a given system, which cells represent distinct lattices.

4.4.2 Monoclinic lattices

The symmetry at each point in a monoclinic lattice is $\frac{2}{m}$. The choice of unit cell leads to the conditions $a \not\subset b \not\subset c$ and $\alpha = \gamma = 90° \not\subset \beta$, with β obtuse. The unique β angle corresponds to the characteristic two-fold symmetry axis being parallel to the y axis, the normal convention.

Fig. 4.8 is a stereoview of a *P* monoclinic unit cell. It is necessary to consider next

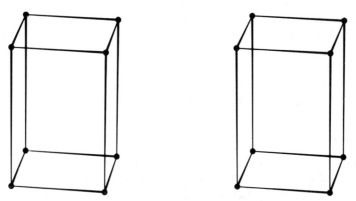

Fig. 4.8—Stereoview of a *P* unit cell in a monoclinic lattice.

if any valid centred monoclinic unit cell represents a new *arrangement* of points within the same crystal system; we can approach this problem in a geometrical manner.

In three dimensions, a unit cell may be centred on any pair of opposite faces, on all faces or at its geometrical (body) centre. These combinations lead to the notation for lattice unit cells given in Table 4.3.

Table 4.3—Lattice unit-cell notation

Centring site/s in unit cell	Symbol	Miller indices of centred faces	Fractional coordinates of lattice points unique to the unit cell (number of lattice points per unit cell)
None	P	—	$0, 0, 0$
bc faces	A	(100)	$0, 0, 0; 0, \frac{1}{2}, \frac{1}{2}$
ca faces	B	(010)	$0, 0, 0; \frac{1}{2}, 0, \frac{1}{2}$
ab faces	C	(001)	$0, 0, 0; \frac{1}{2}, \frac{1}{2}, 0$
Body	I	—	$0, 0, 0; \frac{1}{2}, \frac{1}{2}, \frac{1}{2}$
All faces	F	(100), (010) (001)	$0, 0, 0; 0, \frac{1}{2}, \frac{1}{2}$ $\frac{1}{2}, 0, \frac{1}{2}; \frac{1}{2}, \frac{1}{2}, 0$

Table 4.3 contains *fractional* coordinates of the lattice points that may be taken to belong to one and the same unit cell (remember that the unit cell is a representative portion of the lattice). They are given in triplets, in the order x, y, z, such that

$$x = X/a \qquad (4.6)$$

$$y = Y/b \qquad (4.7)$$

$$z = Z/c \qquad (4.8)$$

where X, Y and Z are the coordinates of a point in absolute measure, and a, b, and c are the corresponding unit cell dimensions in the same units. Thus, fractional coordinates are dimensionless, and independent of the size and shape of the unit cell.

A monoclinic unit cell centred on the C faces is shown in Fig. 4.9. An A centred cell does not represent an arrangement different from C because if we interchange the x and z axes (y is the only axis attached to the symmetry direction), A centring

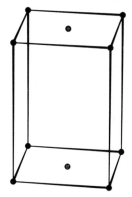 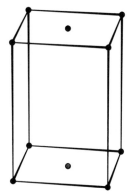

Fig. 4.9—Stereoview of a C unit cell in a monoclinic lattice.

would become C centring. The following equations transform an A to a C unit cell:

$$\mathbf{a}_C = \mathbf{c}_A \tag{4.9}$$

$$\mathbf{b}_C = -\mathbf{b}_A \tag{4.10}$$

$$\mathbf{c}_C = \mathbf{a}_A \tag{4.11}$$

The symmetry axis is still along b (or y), but the change of sign is needed so as to preserve both the right-handed nature of the axes and the obtuse value for the β angle.

Centring on the B faces is illustrated in Fig. 4.10; two unit cells are shown. It is evident that a new unit cell may be defined by the equations

$$\mathbf{a}' = \mathbf{a} \tag{4.12}$$

$$\mathbf{b}' = \mathbf{b} \tag{4.13}$$

$$\mathbf{c}' = -\mathbf{a}/2 + \mathbf{c}/2 \tag{4.14}$$

Here, we have $a' \not\subset b' \not\subset c'$. Since c' lies in the ac plane, $\alpha' = \gamma' = 90°$, but β' is still of a general value, as permitted by the monoclinic symmetry. The new unit cell is P, and of one-half the volume of the B cell (count the numbers of lattice points *per unit cell*), so it is the more conventional choice. Symbolically, $B \equiv P$ in the monoclinic system.

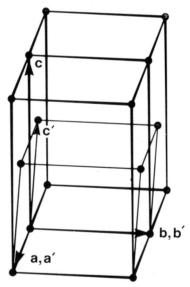

Fig. 4.10—Monoclinic lattice: two B unit cells are drawn, and the conventional P unit cell is shown.

A different result is obtained if the C unit cell is transformed to P in a similar manner (Fig. 4.11):

$$\mathbf{a}' = \mathbf{a}/2 - \mathbf{b}/2 \tag{4.15}$$

$$\mathbf{b}' = \mathbf{b} \tag{4.16}$$

$$\mathbf{c}' = \mathbf{c} \tag{4.17}$$

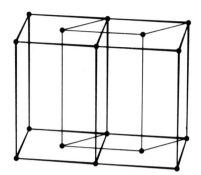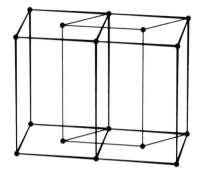

Fig. 4.11—Stereoview of a monoclinic lattice: two C unit cells are drawn; the P unit cell is not the conventional choice.

We have now, $a' \not\subset b' \not\subset c'$ and $\alpha' = 90°$. However, since the ab plane is inclined to the bc plane, γ' (as well as β') is different from $90°$. These results point to the fact that monoclinic C is different from monoclinic P, so that both descriptions are relevant to this system.

Monoclinic I and F unit cells transform to C unit cells in a manner similar to that described above; we shall leave further discussion of these particular transformations until the Problems section at the end of this chapter.

It should be noted that the transformation equations given may not be the only ones possible, but it is desirable to retain the standard conventions of right-handed axes, b direction unique and obtuse β angle for this system.

Transformations can be a practical problem wherein numerical answers are required. The most straightforward method utilises the vector equations of the transformation. Consider the transformation $B \to P$, (4.12)–(4.14); we need to evaluate c' and β'. From (4.14)

$$\mathbf{c}' \cdot \mathbf{c}' = (-\mathbf{a}/2 + \mathbf{c}/2) \cdot (-\mathbf{a}/2 + \mathbf{c}/2) \tag{4.18}$$

whence

$$c'^2 = a^2/4 + c^2/4 - ac(\cos \widehat{\mathbf{ac}})/2 \tag{4.19}$$

so that

$$c' = (a^2/4 + b^2/4 - ac(\cos \beta)/2)^{1/2} \tag{4.20}$$

Next, we have

$$\mathbf{a}' \cdot \mathbf{c}' = a'c' \cos \beta' \tag{4.21}$$

so that

$$\cos \beta' = \mathbf{a} \cdot (-\mathbf{a}/2 + \mathbf{c}/2)/ac' \tag{4.21}$$

$$= (-a^2/2 + ac(\cos \beta)/2)/ac' \tag{4.22}$$

and

$$\beta' = \cos^{-1}[(-a + c \cos \beta)/2c'] \tag{4.23}$$

where c' is given by (4.20). Vector methods are recommended in dealing with these and similar problems. A summary of some simple vector methods is given in Appendix 6.

4.4.3 Orthorhombic lattices

The symmetry at each point in an orthorhombic lattice is *mmm*: the symmetry at lattice points is always that of the highest Laue group of the crystal system (see Section 3.6.5).

The conventional orthorhombic unit cell has the conditions $a \not{C} b \not{C} c$ and $\alpha = \beta = \gamma = 90°$. Figs 4.12 to 4.15 show stereoviews of orthorhombic *P*, *C*, *I* and *F* unit cells. These unit cells are distinct, and correspond to four different arrangements of points which conform to orthorhombic symmetry. The *C* cell could also be called *A* or *B*, but *C* is the standard descriptor. That these four cells do represent different arrangements of points may be demonstrated by the following procedure, used implicitly with the monoclinic lattices (Section 4.4.2).

After centring a *P* unit cell in a given system, consider the following questions, in order:

(a) Does the centred cell represent a true lattice?
(b) If it is a true lattice, is the symmetry different from that in the *P* cell?
(c) If the symmetry is unchanged is the lattice different in type (arrangement of points) from the lattice or lattices already determined for the system? It may be judged by comparing the constants of the reduced unit cell with those characteristic for the system, which is equivalent to asking if the unit cell has been chosen according to convention.

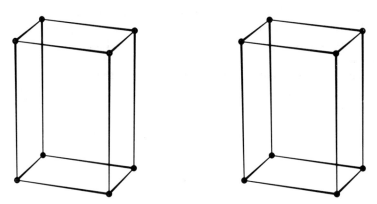

Fig. 4.12—Stereoview of a *P* unit cell in an orthorhombic lattice.

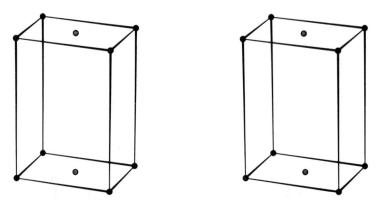

Fig. 4.13—Stereoview of a *C* unit cell in an orthorhombic lattice.

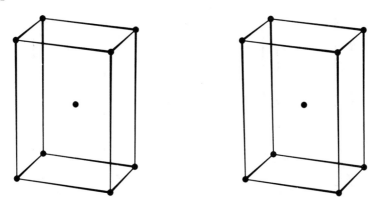

Fig. 4.14—Stereoview of an *I* unit cell in an orthorhombic lattice.

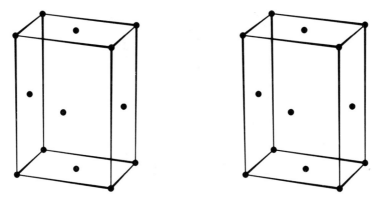

Fig. 4.15—Stereoview of an *F* unit cell in an orthorhombic lattice.

4.4.4 Tetragonal lattices

There are two tetragonal lattices, symmetry $\frac{4}{m}mm$ at each point, represented by *P* and *I* unit cells (Figs 4.16 and 4.17). For each of them, we have the conditions $a = b \,\mathbb{C}\, c$ and $\alpha = \beta = \gamma = 90°$. It is straightforward to show that $C \equiv P$ (Fig. 4.18). If

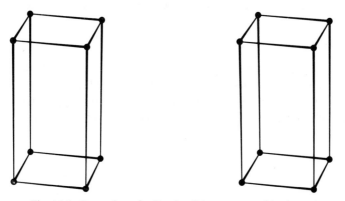

Fig. 4.16—Stereoview of a *P* unit cell in a tetragonal lattice.

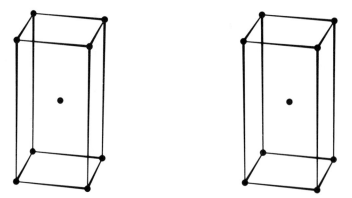

Fig. 4.17—Stereoview of an *I* unit cell in a tetragonal lattice.

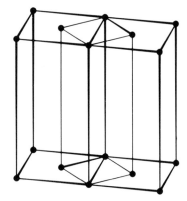

Fig. 4.18—Tetragonal lattice, showing the equivalence of the *P* (conventional) and *C* descriptors.

the *B* faces are centred, the unit cell (and the lattice) no longer has the four-fold symmetry that is characteristic of the tetragonal system. The symmetry is restored, apparently, by centring the *A* faces as well. But this cell does not represent a lattice: the points p_1 and p_2 are not in identical vector environments (Fig. 4.19). Centring the *C* faces in addition to the *A* and *B* produces a tetragonal *F* unit cell, but this is equivalent to *I* (Fig. 4.20). This sequence of changes is shown in projection in Fig. 4.21.

4.4.5 Cubic lattices

There are three cubic lattices, described by the unit cells *P*, *I* and *F*, that are consistent with the symmetry *m3m* at each lattice point (Figs 4.22–4.24). Fig. 4.25 shows that a cubic *P* unit cell after centring on the *A* faces is reduced in symmetry, and represents an orthorhombic lattice (*m3m* → *mmm*). If a cubic *I* unit cell is transformed by the equations

$$\mathbf{a'} = \mathbf{a} \tag{4.24}$$

$$\mathbf{b'} = \mathbf{b} \tag{4.25}$$

$$\mathbf{c'} = \mathbf{a} + \mathbf{c} \tag{4.26}$$

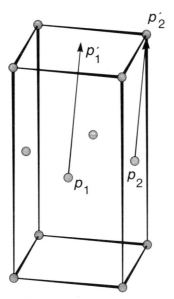

Fig. 4.19—'Tetragonal' unit cell centred on the A and B faces does not represent a lattice.

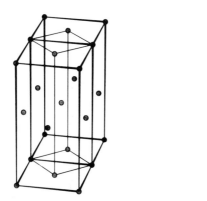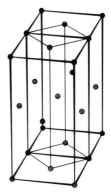

Fig. 4.20—Stereoview of a tetragonal lattice, showing the equivalence of I (conventional) and F descriptors.

the new cell is A centred, but it is still of cubic symmetry because it represents the same lattice: the symmetry is inherent in the non-trivial relationships $a' = b'$, $c' = a'/2$, $\alpha' = \gamma' = 90°$ and $\beta' = 45°$.

4.4.6 Hexagonal lattice

The symmetry at each point in a hexagonal lattice is $\dfrac{6}{m}\,mm$. In the conventional P unit cell, we have $a = b \, ⊄ \, c$, $\alpha = \beta = 90°$ and $\gamma = 120°$. There is only one hexagonal lattice (Fig. 4.26).

Fig. 4.21—Tetragonal lattice in projection on (001): (a) *P* unit cell, (b) *B* unit cell—symmetry reduced to orthorhombic, (c) (*A* + *B*) unit cell—not a true lattice, although symmetry apparently restored, (d) *F* unit cell, equivalent to *I*.

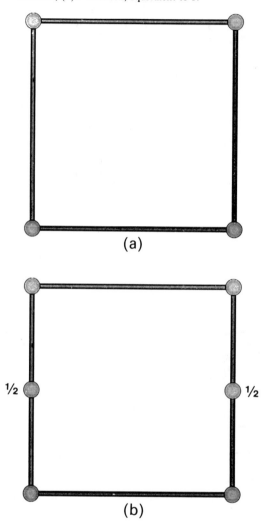

(a)

½   ½

(b)

4.4.7 Trigonal lattices

The two-dimensional hexagonal unit cell in which $a = b$ and $\gamma = 120°$ is compatible with either three-fold or six-fold symmetry (Table 3.1). Thus, the hexagonal *P* unit cell (Fig. 4.26) may be used for certain lattices of trigonal symmetry. However, if the lattice has three-fold symmetry axes, parallel to *z* and passing through the points $\frac{2}{3}, \frac{1}{3}$ and $\frac{1}{3}, \frac{2}{3}$ in the *ab* plane, we have a triply primitive hexagonal unit cell (Fig. 4.27). Thus, some trigonal crystals will have a true *P* unit cell, but others will have a triply primitive hexagonal unit cell. This cell can be transformed to a primitive *R* rhombohedral unit cell, in which $a = b = c$ and $\alpha = \beta = \gamma \not\subset 90°$, $< 120°$ (the unique angle is chosen to be less than 120° so that the volume of the unit cell from equation (A6.26) is real). In this *R* unit cell the three-flod axis is along [111], and it may be thought of as a cube extended along a triad axis.

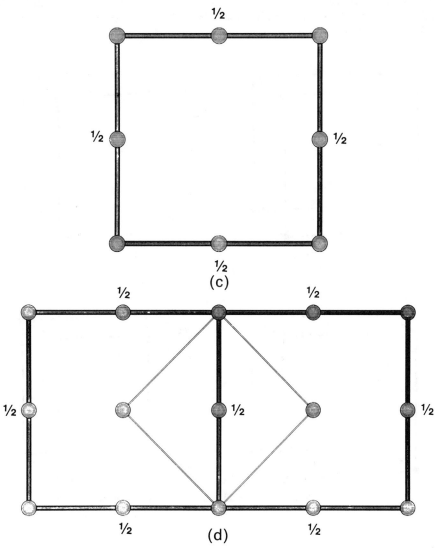

The transformation from the triple hexagonal cell to the rhombohedral unit cell is given by

$$\mathbf{a}_R = \quad \tfrac{2}{3}\mathbf{a}_H + \tfrac{1}{3}\mathbf{b}_H + \tfrac{1}{3}\mathbf{c}_H \tag{4.27}$$

$$\mathbf{b}_R = -\tfrac{1}{3}\mathbf{a}_H + \tfrac{1}{3}\mathbf{a}_H + \tfrac{1}{3}\mathbf{c}_H \tag{4.28}$$

$$\mathbf{c}_R = -\tfrac{1}{3}\mathbf{a}_H - \tfrac{2}{3}\mathbf{b}_H + \tfrac{1}{3}\mathbf{c}_H \tag{4.29}$$

and the inverse transformation is

$$\mathbf{a}_H = \mathbf{a}_R - \mathbf{b}_R \tag{4.30}$$

$$\mathbf{b}_H = \mathbf{b}_R - \mathbf{c}_R \tag{4.31}$$

$$\mathbf{c}_H = \mathbf{a}_R + \mathbf{b}_R + \mathbf{c}_R \tag{4.32}$$

In relation to the triple hexagonal unit cell there are two settings of the R cell, the *obverse* setting (as used above) and the *reverse* setting obtained by rotating the rhombohedral unit cell clockwise about c_H by 60°; the obverse setting is the standard.

The Bravais lattice data are summarized in Table 4.4.

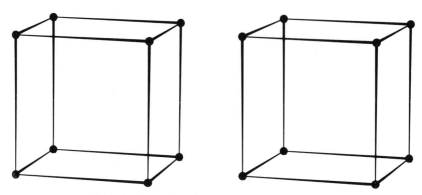

Fig. 4.22—Stereoview of a P unit cell in a cubic lattice.

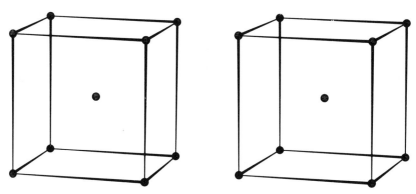

Fig. 4.23—Stereoview of an I unit cell in a cubic lattice.

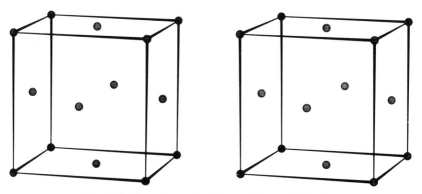

Fig. 4.24—Stereoview of an F unit cell in a cubic lattice.

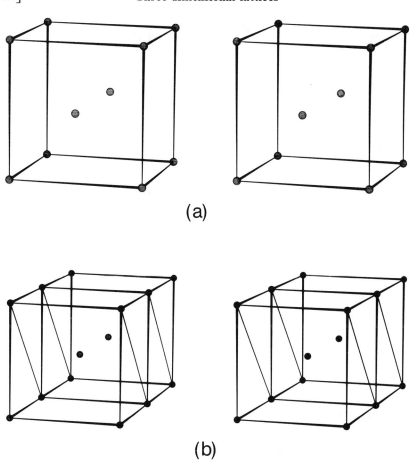

(a)

(b)

Fig. 4.25—Stereoviews of cubic unit cells: (a) $a = b = c$ and $\alpha = \beta = \gamma = 90°$, but centring the A faces alone reduces the symmetry at each point to *mmm* (orthorhombic), (b) $a = b = c$ and $\alpha = \beta = \gamma = 90°$ in a cubic I unit cell; for the cell outlined in thin lines $a' = b' \not{C} c'$, $\gamma' = 90° \not{C} \alpha' \not{C} \beta'$—the symmetry at each point is still *m3m*, but the A cell is unconventional.

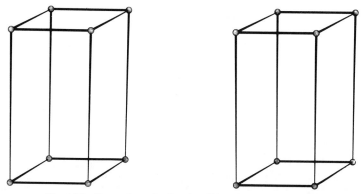

Fig. 4.26—Stereoview of a P unit cell in a hexagonal lattice.

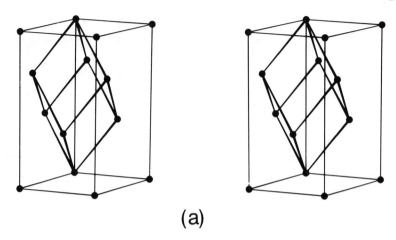

(a)

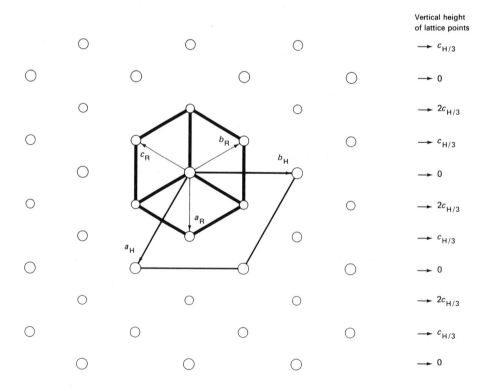

Fig. 4.27—(a) Stereoview of R (primitive) unit cell (obverse setting) and the triply primitive hexagonal unit cell in a trigonal lattice, (b) single view from (a) as seen in projection along the three-fold symmetry axis; the fractions relate to the vertical axis, that is, c in the triply primitive hexagonal unit cell, and [111] in the R unit cell.

Table 4.4—Bravais lattices

System	Unit cell symbols	Axial relationships	Symmetry at each lattice point
Triclinic	P	$a \neq b \neq c$ $\alpha \neq \beta \neq \gamma \neq 90°, 120°$	$\bar{1}$
Monoclinic	P, C	$a \neq b \neq c$ $\alpha = \gamma = 90° \neq \beta$	$\dfrac{2}{m}$
Orthorhombic	P, C, I, F	$a \neq b \neq c$ $\alpha = \beta = \gamma = 90°$	mmm
Tetragonal	P, I	$a = b \neq c$ $\alpha = \beta = \gamma = 90°$	$\dfrac{4}{m} mm$
Cubic	P, I, F	$a = b = c$ $\alpha = \beta = \gamma = 90°$	$m3m$
Hexagonal	P	$a = b \neq c$ $\alpha = \beta = 90°; \gamma = 120°$	$\dfrac{6}{m} mm$
Trigonal (Hexagonal axes)	P	$a = b \neq c$ $\alpha = \beta = 90°; \gamma = 120°$	$\bar{3}m$
Trigonal (Rhombohedral axes)	R	$a = b = c$ $\alpha = \beta = \gamma \neq 90°, <120°$	$\bar{3}m$

4.5 LATTICE DIRECTIONS

Lattice geometry is governed by the three translation vectors **a**, **b** and **c**. It follows from the definition of a lattice that any point may be reached from any other point by performing the basic translations, or positive or negative multiples thereof always in the directions of **a**, **b** or **c**.

Any lattice point may be taken as the origin of the lattice. The vector **r** from the origin to any other lattice point is given by

$$\mathbf{r} = U\mathbf{a} + V\mathbf{b} + W\mathbf{c} \qquad (4.33)$$

where U, V and W are positive or negative integers, or zero, and are the coordinates of the lattice point. Any general position within a unit cell of the lattice has the fractional coordinates x, y, z (see Section 4.4.2). Hence, the vector distance to any from the origin is given by

$$\mathbf{d} = x \cdot \mathbf{a} + y \cdot \mathbf{b} + z \cdot \mathbf{c} \qquad (4.34)$$

The line joining the origin to the lattice points U, V, W; $2U, 2V, 2W$; ... nU, nV, nW defines the *direction*, or directed line, $[UVW]$; the same notation is used for a zone axis (see Section 2.5) because any directon, defined as above, is a possible zone axis since crystal planes are rational (see Section 4.8). A set of directions (or zone axes) related by point-group symmetry defines a form of directions (or zone axes), and the notation $\langle UVW \rangle$ signifies this condition.

The numerical value of r or d in (4.33) or (4.34) may be evaluated from the dot product $\mathbf{r} \cdot \mathbf{r}$ or $\mathbf{d} \cdot \mathbf{d}$ (see Appendix 6).

4.6 RECIPROCAL LATTICE

It is convenient to introduce the reciprocal lattice concept in this chapter. We shall not make use of it, however, until we consider the diffraction of X-rays by crystals in Chapter 6.

For each Bravais lattice, a corresponding reciprocal lattice may be postulated, and it may be derived by the following type of construction.

Consider a monoclinic lattice (P) projected on to the (010) plane (Fig. 4.28(a)). The unit cell is defined by the vectors **a**, **b** and **c**, and families of parallel, equidistant planes may be described through the points of the Bravais lattice. From the origin O of the lattice, lines are constructed normal to the families of planes in the Bravais lattice. (It should be noted that, in general, $[UVW]$ does not coincide with the normal to the *plane* of the same Miller indices.) Along each of these lines, reciprocal lattice points are marked off such that the distance of the first of these points from the origin is inversely proportional to the corresponding interplanar spacing.

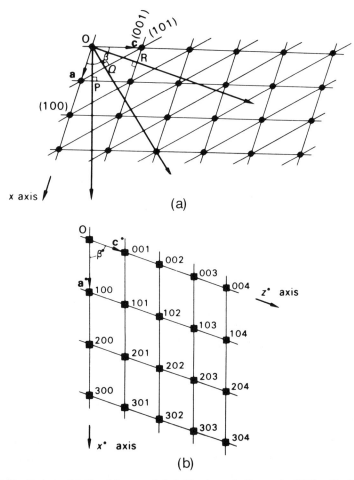

Fig. 4.28—Reciprocal lattice (a) monoclinic lattice in projection on to (010), with P unit cells and traces of the (100), (001) and (101) families of planes shown; the normals to (100), (001) and (101) are OP, OR and OQ respectively, (b) monoclinic reciprocal lattice constructed from (a), in projection on to the a^*c^* plane.

Thus, in the figure, we show the families of planes (100), (001) and (101), and the distances from the origin to the reciprocal lattice points 100, 001 and 101 are proportional to $1/d(100)$, $1/d(001)$ and $1/d(101)$ respectively, where $d(100) = OP$, $d(001) = OQ$ and $d(101) = OR$.

In the context of X-ray diffraction, other families of planes, such as (200), (003) and (203) and, in general, (ph, qk, rl), where p, q and r are integers, are recognized; their reciprocal lattice points are constructed in a similar manner. For any set of planes, such as $(h01)$, the corresponding reciprocal lattice points move further away from the origin, along the $h01$ row, in proportion to the value of h.

In general, we have

$$d^*(hkl) = K/d(hkl) \tag{4.35}$$

where $d(hkl)$ is the distance of the hklth reciprocal lattice point from the origin. The constant K may be chosen as unity or, in the case of practical X-ray diffraction, the wavelength λ of the radiation; in this discussion, we shall take K to be unity.

The reciprocal lattice vectors $\mathbf{d}^*(100)$, $\mathbf{d}^*(010)$ and $\mathbf{d}^*(001)$ may be used to define $\mathbf{a}^*$, $\mathbf{b}^*$ and $\mathbf{c}^*$, the vectors defining the unit cell in the reciprocal lattice. We determine next some equations of importance in the monoclinic reciprocal lattice. From (4.35), we obtain

$$d^*(100) = 1/d(100) = a^* \tag{4.36}$$

But $d(100)$ is, from Fig. 4.28(a), $a \sin \beta$. Hence,

$$a^* = \frac{1}{a \sin \beta} \tag{4.37}$$

Similarly,

$$c^* = \frac{1}{c \sin \beta} \tag{4.38}$$

but

$$b^* = 1/b \tag{4.39}$$

since $d^*(010)$ is normal to the ac plane. The unique angle β^* is given by

$$\beta^* = 180 - \beta \tag{4.40}$$

Next

$$\mathbf{a} \cdot \mathbf{a}^* = aa^* \cos \widehat{\mathbf{a}\mathbf{a}^*} = a \, \frac{1}{a \sin \beta} \cos(\beta - 90°) = 1 \tag{4.41}$$

and similarly for $\mathbf{b} \cdot \mathbf{b}^*$ and $\mathbf{c} \cdot \mathbf{c}^*$. For the mixed products, we have

$$\mathbf{a} \cdot \mathbf{b}^* = ab^* \cos \widehat{\mathbf{a}\mathbf{b}^*} = 0 \tag{4.42}$$

and similarly for the other mixed products.

Relations such as $\mathbf{a} \cdot \mathbf{a}^* = 1$ and $\mathbf{a} \cdot \mathbf{b}^* = 0$ apply in all crystal systems. The general equations for the reciprocal lattice unit cell constants are

$$a^* = \frac{bc \sin \alpha}{V} \tag{4.43}$$

and

$$\cos \alpha^* = \frac{\cos \beta \cos \gamma - \cos \alpha}{\sin \beta \sin \gamma} \tag{4.44}$$

and the cyclic permutations thereof; V is the volume of the Bravais lattice unit cell.

It remains to show that the reciprocal lattice points, as constructed, form a true lattice. For any plane (hkl), from (4.35) with K equal to unity

$$\mathbf{d}(hkl) \cdot \mathbf{d}^*(hkl) = 1 \tag{4.45}$$

We have

$$\frac{d(hkl)}{a/h} = \cos \chi \tag{4.46}$$

where $\cos \chi$ is the direction cosine of $d(hkl)$ (and a^*) with respect to the x axis (see Appendix 4). Also

$$\mathbf{d}(hkl) \cdot \mathbf{a}/h = (d(hkl)a/h) \cos \chi = d^2(hkl) \tag{4.47}$$

Using (4.35)

$$\mathbf{d}(hkl) = \mathbf{d}^*(hkl)/d^{*2}(hkl) \tag{4.48}$$

since $\mathbf{d}(hkl)$ and $\mathbf{d}^*(hkl)$ are co-linear. Therefore

$$\frac{\mathbf{d}^*(hkl)}{d^{*2}(hkl)} \cdot \mathbf{a}/h = \mathbf{d}(hkl) \cdot \mathbf{a}/h = 1/d^{*2}(hkl) \tag{4.49}$$

or

$$\mathbf{d}^*(hkl) \cdot \mathbf{a} = h \tag{4.50}$$

Similarly

$$\mathbf{d}^*(hkl) \cdot \mathbf{b} = k \tag{4.51}$$

and

$$\mathbf{d}^*(hkl) \cdot \mathbf{c} = l \tag{4.52}$$

whence

$$\mathbf{d}^*(hkl) = h\mathbf{a}^* + k\mathbf{b}^* + l\mathbf{c}^* \tag{4.53}$$

Hence, the reciprocal lattice points hkl form a true lattice (compare (4.33)), and any point hkl in the reciprocal lattice represents a family of planes in the Bravais lattice.

The reciprocal lattice has the same symmetry as the Bravais lattice from which it was deduced. This fact may be appreciated from a comparison of the constructions of the reciprocal lattice and the stereogram. Both of these constructions are built up from normals to planes, so that the symmetry expressed through the poles of a stereogram is the same as that at the reciprocal lattice points.

4.7 GENERAL TRANSFORMATIONS

In the discussion of lattices, we were concerned with the transformations of the vectors which specify different unit cells in a given lattice. A transformation of the unit cell produces concomitant changes in related parameters, and we need to consider how they, too, transform. In practical crystallography, we may not, at first, choose the best unit cell, so that subsequent transformations become a necessity.

4.7.1 Transformations without change of origin of the unit cell

The parameters that are most often involved in unit cell transformations are the following:

(a) Bravais lattice unit cell vectors $\quad\quad\quad\quad$ **a, b, c**
(b) Zone symbols and directions $\quad\quad\quad\quad$ *U, V, W*
(c) Fractional coordinates of points in the unit cell $\quad$ *x, y, z*
(d) Miller indices $\quad\quad\quad\quad\quad\quad\quad\quad$ *h, k, l*
(e) Reciprocal lattice unit cell vectors $\quad\quad\quad$ **a*, b*, c***

Bravais lattice unit cell vectors
We have considered several transformations of unit cell vectors. The results may be written generally in the following manner. Let **a**, **b** and **c** be transformed to **a′**, **b′** and **c′**, such that

$$\mathbf{a'} = s_{11}\mathbf{a} + s_{12}\mathbf{b} + s_{13}\mathbf{c} \tag{4.54}$$

$$\mathbf{b'} = s_{21}\mathbf{a} + s_{22}\mathbf{b} + s_{23}\mathbf{c} \tag{4.55}$$

$$\mathbf{c'} = s_{31}\mathbf{a} + s_{32}\mathbf{c} + s_{33}\mathbf{c} \tag{4.56}$$

Following the notation in Section 3.9, we have

$$\begin{vmatrix} \mathbf{a'} \\ \mathbf{b'} \\ \mathbf{c'} \end{vmatrix} = \begin{vmatrix} s_{11} & s_{12} & s_{13} \\ s_{21} & s_{22} & s_{23} \\ s_{31} & s_{32} & s_{33} \end{vmatrix} \cdot \begin{vmatrix} \mathbf{a} \\ \mathbf{b} \\ \mathbf{c} \end{vmatrix} \tag{4.57}$$

or concisely

$$\mathbf{a'} = \mathbf{S} \cdot \mathbf{a} \tag{4.58}$$

where **a** and **a′** represent the two sets of vectors **a, b, c** and **a′, b′, c′** and **S** is the matrix of the elements s_{ij}. The inverse transformation may be written

$$\mathbf{a} = \mathbf{S}^{-1} \cdot \mathbf{a'} \tag{4.59}$$

where $\mathbf{S}^{-1}$ is another matrix, the inverse of **S**, represented by

$$\mathbf{S}^{-1} = \begin{vmatrix} t_{11} & t_{12} & t_{13} \\ t_{21} & t_{22} & t_{23} \\ t_{31} & t_{32} & t_{33} \end{vmatrix} \tag{4.60}$$

The elements t_{ij} may be obtained by rearranging (4.54)–(4.56), or by the following method:

$$t_{ij} = (-1)^{i+j} |\mathbf{M}_{ji}| / |\mathbf{S}| \tag{4.61}$$

where $|\mathbf{M}_{ji}|$ is the minor determinant of elements s_{ij}, and is obtained by striking out the jth row and ith column from $\mathbf{S}$; $|\mathbf{S}|$ is the determinant of $\mathbf{S}$. As an example, let us consider (4.27)–(4.29) and (4.30)–(4.32). We can write

$$\mathbf{a}_R = \mathbf{S} \cdot \mathbf{a}_H \tag{4.62}$$

where $\mathbf{S}$ is the matrix

$$\begin{bmatrix} \frac{2}{3} & \frac{1}{3} & \frac{1}{3} \\ -\frac{1}{3} & \frac{1}{3} & \frac{1}{3} \\ -\frac{1}{3} & -\frac{2}{3} & \frac{1}{3} \end{bmatrix} \tag{4.63}$$

The value of $|\mathbf{S}|$ may be evaluated by the cross-multiplication rule:

$$\tag{4.64}$$

whence

$$|\mathbf{S}| = 2/27 - 1/27 + 2/27 + 1/27 + 4/27 + 1/27 = 1/3 \tag{4.65}$$

$$|\mathbf{M}_{11}| = \begin{vmatrix} \frac{1}{3} & \frac{1}{3} \\ -\frac{2}{3} & \frac{1}{3} \end{vmatrix} = 1/9 + 2/9 = 1/3 \tag{4.66}$$

and

$$t_{11} = (-1)^2(1/3)/(1/3) = 1 \tag{4.67}$$

Similarly

$$t_{21} = (-1)^3(0)/(1/3) = 0 \tag{4.68}$$

and

$$t_{31} = (-1)^4(1/3)/(1/3) = 1 \tag{4.69}$$

Continuing in this way, we can build up the matrix

$$\mathbf{S}^{-1} = \begin{bmatrix} 1 & \bar{1} & 0 \\ 0 & 1 & \bar{1} \\ 1 & 1 & 1 \end{bmatrix} \tag{4.70}$$

so that we have

$$\mathbf{a}_H = \mathbf{S}^{-1} \cdot \mathbf{a}_R \tag{4.71}$$

which is a matrix expression of (4.30)–(4.32). In some examples, it may be easier to write down the inverse matrix by inspection of the equations for the forward transformation.

Zone symbols and directions

From Section 4.5, we have

$$\mathbf{r} = U\mathbf{a} + V\mathbf{b} + W\mathbf{c} \tag{4.72}$$

and for the transformed unit cell

$$\mathbf{r} = U'\mathbf{a}' + V'\mathbf{b}' + W'\mathbf{c}' \tag{4.73}$$

Through (4.72) and (4.59), we now obtain

$$\mathbf{r} = \boxed{U\ V\ W} \cdot \begin{vmatrix} \mathbf{a} \\ \mathbf{b} \\ \mathbf{c} \end{vmatrix} = \boxed{U\ V\ W} \cdot \mathbf{S}^{-1} \cdot \begin{vmatrix} \mathbf{a}' \\ \mathbf{b}' \\ \mathbf{c}' \end{vmatrix} \tag{4.74}$$

But from (4.73)

$$\mathbf{r} = \boxed{U'\ V'\ W'} \cdot \begin{vmatrix} \mathbf{a}' \\ \mathbf{b}' \\ \mathbf{c}' \end{vmatrix} \tag{4.75}$$

whence

$$\boxed{U'\ V'\ W'} = \boxed{U\ V\ W} \cdot \mathbf{S}^{-1} \tag{4.76}$$

By transposition

$$\begin{vmatrix} U' \\ V' \\ W' \end{vmatrix} = \begin{vmatrix} U \\ V \\ W \end{vmatrix} \cdot (\mathbf{S}^{-1})^{\mathsf{T}} \tag{4.77}$$

or more concisely

$$\mathbf{U}' = (\mathbf{S}^{-1})^{\mathsf{T}} \cdot \mathbf{U} \tag{4.78}$$

where $\mathbf{U}$ and $\mathbf{U}'$ refer to the appropriate triplet of zone symbols.

From (4.60), it follows that

$$(\mathbf{S}^{-1})^{\mathsf{T}} = \begin{vmatrix} t_{11} & t_{21} & t_{31} \\ t_{12} & t_{22} & t_{32} \\ t_{13} & t_{23} & t_{33} \end{vmatrix} \tag{4.79}$$

Similarly, we can show that

$$\mathbf{U} = \mathbf{S}^{\mathsf{T}} \cdot \mathbf{U}' \tag{4.80}$$

Coordination of points in the unit cell

For any point x, y, z, a vector $\mathbf{r}$ from the origin to the point is

$$\mathbf{r} = x\mathbf{a} + y\mathbf{b} + z\mathbf{c} \tag{4.81}$$

Comparison of (4.72) and (4.81) shows that coordinates transform as do zone symbols. Thus,

$$x' = (S^{-1})^T \cdot x \tag{4.82}$$

Miller indices

From (4.53) and (4.72)

$$d^*(hkl) \cdot r = (ha^* + kb^* + lc^*) \cdot (Ua + Vb + Wc) = hU + kV + lW \tag{4.83}$$

Thus, with (4.80)

$$d^*(hkl) \cdot r = \begin{vmatrix} h & k & l \end{vmatrix} \cdot \begin{vmatrix} U \\ V \\ W \end{vmatrix} = \begin{vmatrix} h & k & l \end{vmatrix} \cdot S^T \cdot \begin{vmatrix} U' \\ V' \\ W' \end{vmatrix} \tag{4.84}$$

But

$$d^*(h'k'l') \cdot r = \begin{vmatrix} h' & k' & l' \end{vmatrix} \cdot \begin{vmatrix} U' \\ V' \\ W' \end{vmatrix} \tag{4.85}$$

because $d^*(hkl)$ and $d^*(h'k'l')$ are one and the same. Hence

$$\begin{vmatrix} h' & k' & l' \end{vmatrix} = \begin{vmatrix} h & k & l \end{vmatrix} \cdot S^T \tag{4.86}$$

Transposing

$$\begin{vmatrix} h' \\ k' \\ l' \end{vmatrix} = \begin{vmatrix} h \\ k \\ l \end{vmatrix} \cdot S \tag{4.87}$$

or

$$h' = S \cdot h \tag{4.88}$$

where h and h' represent h, k, l and h', k', l', so that Miller indices transform in the same way as do unit cell vectors.

If we operate on both sides of (4.88) by S^{-1}, we obtain

$$S^{-1} \cdot h' = S^{-1} \cdot S \cdot h \tag{4.89}$$

Since, for any matrix S

$$S \cdot S^{-1} = S^{-1} \cdot S = 1 \tag{4.90}$$

we have

$$h = S^{-1} \cdot h' \tag{4.91}$$

Reciprocal unit cell vectors
From (4.53), we can obtain

$$\mathbf{d}^*(hkl) = \begin{array}{|ccc|} \mathbf{a}^* & \mathbf{b}^* & \mathbf{c}^* \end{array} \cdot \begin{array}{|c|} h \\ k \\ l \end{array} \tag{4.92}$$

Thus,

$$\mathbf{d}^*(hkl) = \begin{array}{|ccc|} \mathbf{a}^* & \mathbf{b}^* & \mathbf{c}^* \end{array} \cdot \mathbf{S}^{-1} \cdot \begin{array}{|c|} h' \\ k' \\ l' \end{array} \tag{4.93}$$

In the new reciprocal unit cell

$$\mathbf{d}^*(hkl) = \begin{array}{|ccc|} \mathbf{a}'^* & \mathbf{b}'^* & \mathbf{c}'^* \end{array} \cdot \begin{array}{|c|} h' \\ k' \\ l' \end{array} \tag{4.94}$$

so that

$$\begin{array}{|ccc|} \mathbf{a}'^* & \mathbf{b}'^* & \mathbf{c}'^* \end{array} = \begin{array}{|ccc|} \mathbf{a}^* & \mathbf{b}^* & \mathbf{c}^* \end{array} \cdot \mathbf{S}^{-1} \tag{4.95}$$

Transposing

$$\begin{array}{|c|} \mathbf{a}'^* \\ \mathbf{b}'^* \\ \mathbf{c}'^* \end{array} = \begin{array}{|c|} \mathbf{a}^* \\ \mathbf{b}^* \\ \mathbf{c}^* \end{array} \cdot (\mathbf{S}^{-1})^{\mathsf{T}} \tag{4.96}$$

or

$$\mathbf{a}'^* = (\mathbf{S}^{-1})^{\mathsf{T}} \cdot \mathbf{a}^* \tag{4.97}$$

so that the reciprocal unit cell vectors transform as do zone symbols and coordinates of points in the Bravais unit cell.

As an example application, from (4.71) we have

$$\mathbf{a}_H = \mathbf{S}^{-1} \cdot \mathbf{a}_R \tag{4.98}$$

where $\mathbf{S}^{-1}$ is given by (4.70). It follows that

$$h_H = h_R - k_R \tag{4.99}$$

$$k_H = k_R - l_R \tag{4.100}$$

$$l_H = h_R + k_R + l_R \tag{4.101}$$

Hence

$$(-h + k + l)_H = 3k_R \tag{4.102}$$

Consequently, if a trigonal (rhombohedral) crystal is indexed on hexagonal (Miller–Bravais) axes, the sum $(-h + k + l)$ will be a multiple of three.

From (A6.16), it follows that the ratio of the volumes of the transformed unit cell (V') and the original unit cell (V) are given by

$$V'/V = |\mathbf{S}| = 1/|\mathbf{S}^{-1}| \qquad (4.103)$$

where $\mathbf{S}$ is the matrix transforming the original unit cell to the new unit cell.

Thus, for (4.98) to (4.101), we have $V_H/V_R = 3$, that is, the hexagonal unit cell is three times the volume of the rhombohedral unit cell. The same result is obtained, of course, by the use of (A6.26), or by counting the numbers of lattice points per unit cell in each case.

4.8 LAW OF RATIONAL INDICES

In a study of crystals, we find that when the faces of a crystal are allocated Miller indices (indexed) in the simplest manner, the Miller indices are small whole numbers; rarely does an index exceed 5. The Law of Rational Indices embodies this result, and we can interpret the law in terms of the lattice theory now developed.

Consider the projection of an orthorhombic lattice, such as that shown by Fig. 4.29. The traces of three families of planes (100), (110), and $(2\bar{3}0)$ are outlined in relation to the P unit cell given. We can show that as the Miller indices increase, the reticular densities (lattice points per unit area) decrease. For example, if the unit cell has the dimensions $a = 0.3$ nm, $b = 0.4$ nm and $c = 0.6$ nm, then the reticular densities of the planes (100), (110) and $(2\bar{3}0)$ are in the ratios 0.050:0.040:0.017.

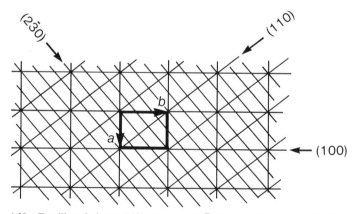

Fig. 4.29—Families of planes (100), (110) and $(2\bar{3}0)$ in an orthorhombic lattice.

In simple crystal structures, the component atoms often occupy positions on lattice points, so that the population of atoms on a given plane may be related directly to the reticular density of lattice points that plane. The faces on a crystal represent the terminations of families of planes, and a crystal grows in such a way that the external faces are planes of highest reticular density. This situation produces a more energetically stable system, because of the better balance of forces, than one in which the surface contains, on the atomic scale, relatively large holes.

As we have shown that the planes of highest reticular density are those of lower Miller indices, the Law of Rational Indices is readily understood. We note also that the Miller indices of crystal planes, or possible planes, are always rational. If the unit cells from Fig. 4.29 packed in the manner shown in Fig. 4.30, the external faces will still have rational Miller indices. A crystal does not build in fractions of a unit cell, so that the apparent steps shown here as being on {110} are only of atomic dimensions.

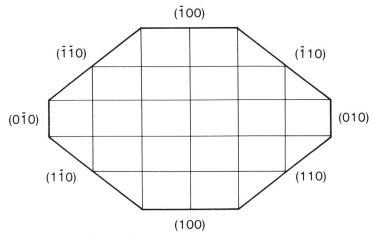

Fig. 4.30—Typical shape for the cross-section of an orthorhombic crystal; the zone axis [001] is normal to the plane of the diagram. The packing of unit cells leads to rational crystal planes, that is, those with integral Miller indices.

4.9 ROTATIONAL SYMMETRY OF CRYSTALS

It is appropriate to show, at this stage, how lattice geometry can be used to account for the permissible rotational symmetries in crystals. In Fig. 4.31, A and B represent two adjacent lattice points in any row of a lattice; their repeat distance is t. An R-fold rotation axis will be considered to act at each point of the lattice, and in a direction that is normal to a plane containing A and B.

An anticlockwise rotation of ϕ about the axis through A maps B on to B', and a similar but clockwise rotation about the axis through B maps A on to A'. The lines $A'S$ and $B'T$ are drawn perpendicular to the line through A and B to meet it in S and T respectively.

Now, the triangles $AB'T$ and $BA'S$ are congruent. Hence,

$$A'B' \text{ is parallel to } AB \tag{4.104}$$

In a lattice, any two points in a row must be separated by an integral multiple of the repeat distance in the direction of that row. Thus, we have

$$A'B' = mt \tag{4.105}$$

where m is an integer. Furthermore,

$$A'B' = t - (AT + BS) = t - 2t(\cos \phi) \tag{4.106}$$

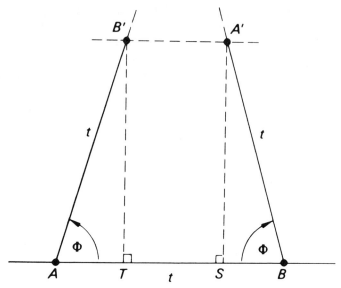

Fig. 4.31—Diagram to show the possible rotational symmetries for crystals based upon lattices.

whence

$$m = 1 - 2 \cos \phi \qquad (4.107)$$

or

$$\cos \phi = (1 - m)/2 = M/2 \qquad (4.108)$$

where M is another integer. Since $|\cos \phi| \leqslant 1$, the only values of M which are admissible are 0, 1, 2, -1 and -2; these values correspond to ϕ of 90°, 60°, 360° (0°), 120° and 180° respectively. Consequently, the rotational symmetries are constrained to the five values of R, 1, 2, 3, 4 and 6. This analysis gives a deeper meaning to the unit cell packing modes considered already in Chapter 3.

PROBLEMS 4

1. Two nets are described by the unit cells (i) $a = b$, $\gamma = 90°$, and (ii) $a = b$, $\gamma = 120°$. In each case (a) what is the symmetry at each lattice point, (b) to which two-dimensional system does each net belong, and (c) what are the results of centring each cell? Give transformation equations as appropriate.

2. A unit cell has the dimensions $a = 0.5$ nm, $b = 0.6$ nm, $c = 0.7$ nm, $\alpha = 90°$, $\beta = 120°$ and $\gamma = 90°$. What crystal system does it represent, and what is the symmetry at each lattice point? Calculate the length of $[31\bar{2}]$.

3. Define the term *Bravais lattice*. Which of the following unit cells represent Bravais lattices? Give reasons for your answers, and write down transformation matrices as appropriate.

   (i) Orthorhombic B
   (ii) Tetragonal A
   (iii) Triclinic I

4. Determine, with the aid of drawings as necessary, the transformation equations for the unit cells in

   (i) Monoclinic $I \to$ Monoclinic C
   (ii) Rhombohedral $F \to$ Rhombohedral R
   (iii) Tetragonal $C \to$ Tetragonal P

5. The relationships $a \not\subset b \not\subset c$, $\alpha \not\subset \beta \not\subset 90°$ and $\gamma = 90°$ may be said to represent a diclinic system. Is it a new crystal system? If so, how so, and if not, why not?

6. Draw a cubic F unit cell and outline a primitive rhombohedral unit cell within it; let the unique axis of the rhombohedron coincide with $[\bar{1}11]$ in the cube. What is the ratio $V(\text{rhombohedron})/V(\text{cube})$?

7. A rhombohedral I unit cell has the dimensions $a_I = 7.00$ Å and $\alpha_I = 50.0°$. Show that the unit cell given by $\mathbf{a}_R = \mathbf{S} \cdot \mathbf{a}_I$, where $\mathbf{S}$ is the matrix

$$
\begin{vmatrix}
-\frac{1}{2} & \frac{1}{2} & \frac{1}{2} \\
\frac{1}{2} & -\frac{1}{2} & \frac{1}{2} \\
\frac{1}{2} & \frac{1}{2} & -\frac{1}{2}
\end{vmatrix},
$$

is a rhombohedron. Calculate the new dimensions a_R and α_R and determine the unit cell type.

8. An orthorhombic unit cell (I) is transformed to another unit cell (II) by the equation $\mathbf{a}_{II} = \mathbf{S} \cdot \mathbf{a}_I$, where $\mathbf{S}$ is the matrix

$$
\begin{vmatrix}
1 & \bar{1} & 0 \\
1 & 1 & 0 \\
0 & 0 & 1
\end{vmatrix}
$$

What is the volume of cell$_{II}$ in terms of cell$_I$? Write down $\mathbf{S}^T$, $\mathbf{S}^{-1}$ and $(\mathbf{S}^{-1})^T$. Hence, or otherwise, determine the coordinates of the point $(0.123, -0.671, 0.314)_I$ when referred to cell$_{II}$ from cell$_I$.

9. Determine the relative reticular densities of the planes (100), (110), ($1\bar{3}0$) and (042) in a cubic I unit cell. Which of these planes might be expected to form the external faces of a simple crystal based on this lattice?

10. The monoclinic unit cell of gypsum, $CaSO_4 \cdot 2H_2O$, has been selected in different ways:

	a/Å	b/Å	c/Å	β/deg
(I)	10.51	15.15	6.545	151°43′
(II)	5.67	15.15	6.545	118°35′
(III)	10.51	15.15	6.28	99°18′

(a) Draw a projection of the Bravais lattice, using either data I or II, on to the plane (010); a scale of 1 cm = 1 Å is suitable. It is convenient to use a transparent guide-sheet, and to draw a grid of lines to locate the lattice points, and then to prick through the points on to another sheet of paper. On the latter sheet, outline the three different unit cells, taking a common origin at the centre of the paper.

(b) If unit cell I is C, what are the unit cell types for II and III? What are the ratios of the volume of cell III to the volumes of cells I and II? Which unit cell would you choose? Give reasons.

(c) Derive from your drawings the matrices needed for all possible transformations of the unit cell vectors among the unit cells given.

(d) Calculate the dimensions of the reciprocal unit cell for the Bravais unit cell that you have chosen in (c). Sketch the Bravais unit cell in projection on to (010), and draw the reciprocal unit cell on the same diagram.

(e) Derive the transformation matrix for reciprocal unit cell I to reciprocal unit cell II. Evaluate a_{II}^{*}, b_{II}^{*}, c_{II}^{*} and β_{II}^{*} from the matrix; check the results with the given experimental data.

(f) Carry out the following transformations with the appropriate matrices:

    (i) $(hkl)_{III}$ from $(13\bar{2})_{I}$

    (ii) $[UVW]_{II}$ from $[2\bar{1}3]_{III}$

    (iii) $(x, y, z)_{III}$ from $(0.6, 0.5, -0.3)_{I}$

5

Space groups

For true understanding, comprehension of detail is
imperative. Since such detail is wellnigh infinite
our knowledge is always superficial and imperfect.

Duc François de la Rochefoucauld, 1665: *Maxims*, No. 160

5.1 INTRODUCTION

We now further our study of the geometry of crystals by considering the symmetries
of extended, ideally infinite, spatial patterns. We recall from Chapter 3 that the
symmetry of any finite body can be described by a point group, and in Chapter 4
we discussed lattices and showed that they provided a mechanism for repetition, to
an infinite extent, by translations parallel to one, two non-collinear or three non-
coplanar directions (reference axes). It is reasonable to ask what the result is of
repeating a motif of an given point-group symmetry by the translations of a lattice.
We shall see that such a combination enables us to build up patterns like the
arrangements of atoms in crystals; such patterns may be described by space groups.

Space groups may be considered in one, two or three dimensions and, for our
purposes, we define a space group as *an infinite set of symmetry elements, the operation
of which leaves the pattern to which it refers indistinguishable from its condition before
the operation*. Thus, a symmetry operation of a space group maps the pattern to
which it refers on to itself. The feature of indistinguishability before and after a
symmetry operation is evident here, as in our earlier discussion on point-group
symmetry.

Space-group operations do not have to leave a point unmoved, and it is permissible
to apply space-group theory to crystals of finite, experimental size because the number
of repeat units is large in comparison to the size of the repeat unit itself. For example,
the dimension a of the face-centred cubic unit cell of sodium chloride is 0.564 nm.
Thus, in a crystal of experimental size ($ca\ 0.2 \times 0.2 \times 0.2\ mm^3$) there are approximately
5×10^{16} unit cells.

5.2 ONE-DIMENSIONAL SPACE GROUPS

From the single one-dimensional lattice and the two one-dimensional point groups,
we can arrive at the two one-dimensional space groups $\not{p}1$ and $\not{p}m$. They are of

limited interest, and are mentioned here mainly for completeness. We may note that in the structure of magnesium fluoride (MgF_2), for example, the magnesium atoms are at the origin of the unit cell and at the fractional coordinates $\frac{1}{2}, \frac{1}{2}, \frac{1}{2}$. Fluorine atoms occupy the positions $x, x, 0,\ \bar{x}, \bar{x}, 0,\ \frac{1}{2} - x, \frac{1}{2} + x, \frac{1}{2}$ and $\frac{1}{2} + x, \frac{1}{2} - x, \frac{1}{2}$, where $x = 0.39$. Thus, the electron density maxima in the unit cell will coincide with these positions. If we project the density function along the z axis, we obtain the maxima at $0, 0$ and $\frac{1}{2}, \frac{1}{2}$ for magnesium and $x, x,\ \bar{x}, \bar{x},\ \frac{1}{2} - x, \frac{1}{2} + x$ and $\frac{1}{2} + x, \frac{1}{2} - x$ for fluorine. If we now project further, into the one dimension x, we have magnesium at 0 and $\frac{1}{2}$, with fluorine at $x,\ \bar{x},\ \frac{1}{2} - x$ and $\frac{1}{2} + x$. The function can be represented as in Fig. 5.1. The symmetry of the repeating electron density $\rho(x)$ can be described by the one-dimensional space group $\not p m$, with a projected unit cell length a' equal to $a/2$.

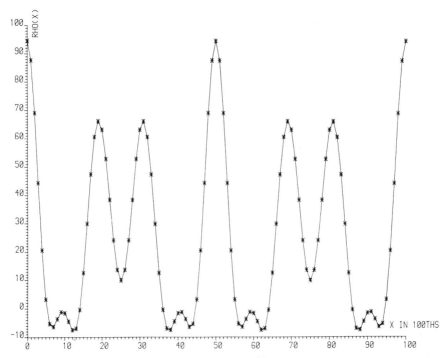

Fig. 5.1—Electron density map for magnesium fluoride MgF_2, projected on to the x axis. The space group is $\not p m$, with a projected unit-cell repeat dimension of $a/2$, that is, $50/100$ along the x axis.

5.3 TWO-DIMENSIONAL SPACE GROUPS (PLANE GROUPS)

Any space group may be considered to be made up of two parts: a pattern motif, and a mechanism for repeating it. We may consider an analogy with a wallpaper, shown with simplicity in Fig. 5.2(a).

The conventional unit cell for this pattern is delineated by the vectors **a** and **b**. The wallpaper can be generated from the unit cell (Fig. 5.2(b)) by repeating it with the translations $n\mathbf{a}$ and $n\mathbf{b}$, where $n = 1, 2, 3 \ldots$ While this technique suffices to build up the pattern, it ignores the symmetry between the two flowers in the unit cell. If flower 1 is reflected across the dashed line, transiently to position 1', and then

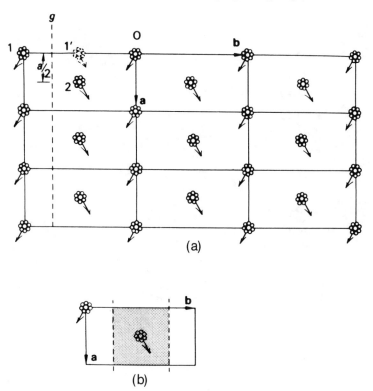

Fig. 5.2—Wallpaper pattern: (a) extended array, with a glide line g inserted, (b) unit cell, showing glide lines and asymmetric unit.

translated parallel to that line by $a/2$, it occupies the position and orientation of flower 2. In other words, the infinite pattern, of which Fig. 5.2(a) is a sample, is mapped on to itself by the symmetry operation. In two dimensions, the symmetry element of this operation is the glide line, symbol g.

The necessary and sufficient pattern unit is a single flower, occupying the asymmetric unit—the shaded (or unshaded) portion of the unit cell in Fig. 5.2(b). If the asymmetric unit is operated on by the space group, pg in this case, the whole pattern is generated. Thus, to apply our analogy, if we have the asymmetric unit for any crystal structure, and we know its space group, we can build up the whole crystal. Fig. 5.3 is another pattern of the same symmetry as that in Fig. 5.2. Can you identify the unit cell and its contents, and the positions of the symmetry elements?

This discussion leads us to a general study of the two-dimensional space groups, or plane groups. There are seventeen plane groups, but we need not consider them all here. They may be classified within the two-dimensional system names listed in Table 3.1.

5.3.1 Plane groups in the oblique system

There are two groups classified here, $p1$ and $p2$; we will consider $p2$ in detail. Fig. 5.4(a) is a motif showing two-fold symmetry (point group 2), and Fig. 5.4(b) is the lattice of the oblique system, divided by a framework into a number of p primitive unit cells;

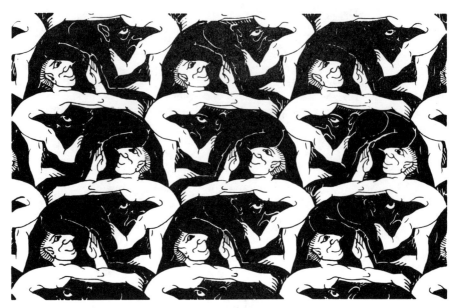

Fig. 5.3—Pattern of plane group *pg*.

the symmetry at each lattice point is 2. The motif may be set down at each lattice point, in a given orientation, so that the two-fold rotation points of the motif and the lattice coincide (Fig. 5.4(c)). Additional two-fold rotation points, which may not have been anticipated, are introduced into the extended pattern of motifs.

In general, we shall not need to draw several unit cells of a space group pattern. One unit cell will suffice, provided that we complete the pattern around the four corners of the unit cell. Fig. 5.5 illustrates a standard drawing for *p*2. By convention, the origin is chosen on a two-fold rotation point, and the *x* axis runs from top to bottom, with the *y* axis running from left to right. Thus, the origin is at the top left-hand corner of the unit cell, but each corner is symmetry-equivalent position. We must always remember that, as with lattices, a unit cell is a representative portion of an infinite array, whether in one, two or three dimensions.

The asymmetric unit, represented by ○, may be an atom, a molecule or part of a molecule (it may or may not be itself symmetrical), and can be placed anywhere in the appropriate fraction of the unit cell, but conveniently near to the origin, and then repeated by the symmetry *p*2 so as to build up the complete pattern, ensuring that the motif around the origin is repeated by the translations **a**, **b** and **a** + **b**. Then, the additional two-fold rotation points can be identified. After some practice, it becomes possible to locate them by inspection. However, at first, they can be determined by taking one point, say *x*, *y*, and considering how every other point in the diagram can be reached from it by *one* symmetry operation, that is, 2 in this group, together with unit cell translations as necessary.

The lists of fractional coordinates in Fig. 5.5 refer to symmetry-related sites in the unit cell. The maximum number of such sites considered to be generated by a space group symmetry constitutes the *general equivalent positions* (sites of symmetry 1). In *p*2 they are *x*, *y* and $\bar{x}$, $\bar{y}$. We could have used $1 - x$, $1 - y$ instead of $\bar{x}$, $\bar{y}$, but it is more usual to work with a set of coordinates near to the origin of the unit cell.

Each line of coordinates lists, in order, the number of positions in the set, the Wyckoff notation for the set, the symmetry at each site in the set, and the fractional coordinates of the sites in the set.

It can happen that the number of entities in the unit cell is less than the number of general equivalent positions. Then, the entities must occupy *special equivalent positions* (sites of symmetry greater than 1) in the space group, and must possess the symmetry of the site, or a symmetry of which the site symmetry is a sub-group. The number of sites in a special set is always a sub-multiple of the number of general equivalent positions. Exceptions to the space-group occupancy rule may occur in disordered structures, where a given set of positions, general or special, may be occupied in a random manner by fewer entities than the space group requires. We shall not, however, be concerned with disordered structures in this book.

In $p2$ there are four sets of special positions, a, b, c and d. Again, it would be possible to choose alternatives, for example, a, b, d, and the site $\frac{1}{2}, 1$ in place of the symmetry-equivalent c. We can show that the four sites a, b, c and d are unique to one unit cell by the shifted-parallelogram technique (Fig. 4.3).

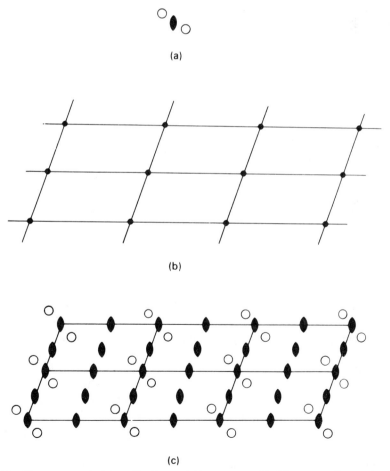

(a)

(b)

(c)

Fig. 5.4—Plane group $p2$: (a) motif with two-fold symmetry, (b) oblique net, with p unit cells outlined; the symmetry at each lattice point is 2, (c) combination of (a) and the repeat mechanism of (b) to give the extended pattern of plane group $p2$.

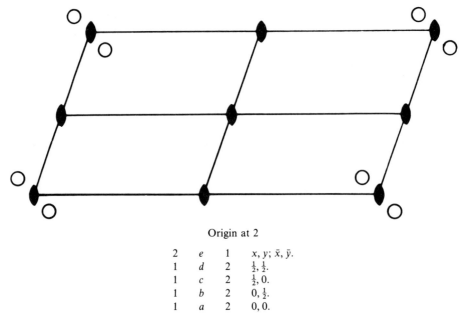

Origin at 2

2	e	1	$x, y; \bar{x}, \bar{y}.$
1	d	2	$\frac{1}{2}, \frac{1}{2}.$
1	c	2	$\frac{1}{2}, 0.$
1	b	2	$0, \frac{1}{2}.$
1	a	2	$0, 0.$

Fig. 5.5—Conventional drawing and description of plane group p2. The lines that divide the unit cell into four quadrants are drawn for convenience. The general equivalent positions are set (e), and sets (d), (c), (b) and (a) are special equivalent positions.

5.3.2 Plane groups in the rectangular system

The rectangular system includes point groups *m* and *2mm*, and has two lattices, represented conventionally by unit cells *p* and *c*. It is not possible to accommodate *m* symmetry in the oblique system. One can always define *two* points with an apparent mirror relationship in a parallelogram, but the arrangement cannot be extended by lattice translations without increasing indefinitely the number of positions in each unit cell.

Let us consider first the plane groups *pm* and *cm* where, by convention (see Table 3.1), the *m* line is normal to the *x* axis (full symbols *pm1* and *cm1*—see Section 5.3.5). These groups may be constructed in a manner similar to that used for *p2* but with an *m* symmetry motif, and we refer immediately to Fig. 5.6. The origin is chosen on *m*, but its *y* coordinate is not defined until the first entity is introduced into the unit cell. In *pm*, there are two general equivalent positions, and two sets of special positions on *m* lines.

Plane group *cm* introduces several new features. The coordinates lists are preceded by the expression $(0, 0; \frac{1}{2}, \frac{1}{2})+$, which means that the translations $0, 0$ (trivial) and $\frac{1}{2}, \frac{1}{2}$ are added to the coordinates listed below them. Thus, the full listing of set *b* in *cm* is $x, y, \bar{x}, \bar{y}, \frac{1}{2}+x, \frac{1}{2}+y$ and $\frac{1}{2}-x, \frac{1}{2}+y$. Given *x*, for example, the positions $\frac{1}{2}+x$ and $\frac{1}{2}-x$ may be obtained by moving *a*/2 from the origin coupled with movements of $+x$ and $-x$ respectively.

Centring the unit cell, in conjunction with *m* symmetry, introduces the glide line *g*, graphic symbol –––. The glide lines interleave the *m* lines, and their operation is a reflexion across the line coupled with a translation of one-half of the repeat

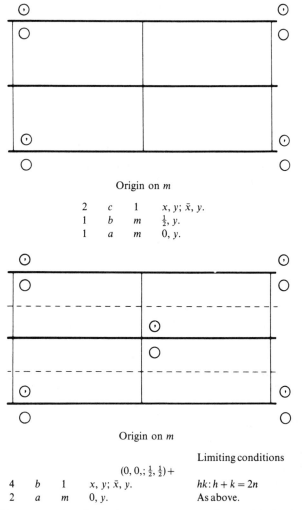

Fig. 5.6—Plane groups based on point group *m*: (a) *pm*, (b) *cm*, with descriptions.

distance in the direction of the line. Thus, symbolically,

$$x, y \xrightarrow{\text{g}\perp x} \tfrac{1}{2} - x, \tfrac{1}{2} + y$$

Glide lines will arise where any centred arrangement exists in the presence of *m* symmetry.

Not all plane groups can be obtained by the combination of a point-group motif and a lattice. Those that can, like *p2*, *pm* and *cm* have been called 'point-space groups' by some authors. We may ask now what is meant by the symbols *pg* and *cg*. The first of these is one of the seventeen plane groups, and is represented by the pattern in Figs 5.2(a) and 5.4. Plane group *cg* is another (non-standard) name for *cm* and is not to be interpreted as a new plane group.

There is only one set of special equivalent positions in *cm*, in contradistinction to the two sets in *pm*, because the centring requires that the two lines† $[0, y]$ and $[\frac{1}{2}, \frac{1}{2} + y]$ be included in one and the same set. If we try to write two sets in *cm*, by analogy with *pm*, we obtain

$$0, y; \tfrac{1}{2}, \tfrac{1}{2} + y \tag{5.1}$$

$$\tfrac{1}{2}, y; 1(\text{or } 0), \tfrac{1}{2} + y \tag{5.2}$$

The expressions (5.1) and (5.2) differ only in the value of the variable *y*, and so do not represent different sets of special positions. We shall not comment upon the columns headed 'Limiting conditions', as in Fig. 5.6(b), until Chapter 6.

Point group *2mm* belongs to the rectangular system. Bearing in mind our discussion on *pm* and *cm*, we may write down eight possible symbols for plane groups related to *2mm*:

$$
\begin{array}{ll}
p2mm^* & c2mm^* \\
p2mg^* & c2mg \\
p2gm & c2gm \\
p2gg^* & c2gg
\end{array}
$$

Those marked with an * are distinct arrangements, and are the standard notations; *p2mg* and *p2gm* differ only by an interchange of the *x* and *y* axes (see Table 3.1); and the four centred groups are identical—each contains interleaving *m* and *g* lines normal to both the *x* and *y* axes. We will consider *p2gg* in detail. When considering any space group, it can be helpful to recall its parent point group. It may be determined simply by removing the unit-cell symbol and then replacing all translational symmetry elements in the symbol by their non-translational counterparts. Thus, *p2gg* is related to *2mm*. Then, we know from Table 3.1 the relative orientations of the symmetry elements in the point-group symbol; and this information is carried over to the space group symbol in a one-to-one correspondence.

In *2mm* the two *m* lines intersect in the two-fold rotation point. Thus, in *p2mm* and *c2mm*, the origin is at *2mm*. When glide symmetry is present, we must not assume that the symmetry lines will intersect in the two-fold rotation point, and mostly they do not.

A correct description of *p2gg* may be reached in more than one way. From the meaning conveyed by the symbol, we know that the first *g* is normal to the *x* axis and the second *g* is normal to *y*. So we may construct a unit cell from the glide lines (Fig. 5.7(a)). Then, we insert a point *x, y* and repeat it by the symmetry *pgg* (Fig. 5.7(b)). We know that the interaction of the two symmetry elements leads to a third (Section 3.6.4), so we may expect to see the positions of the two-fold rotation points (Fig. 5.7(c)). Finally, we can re-orient the diagram with the origin on 2 (Fig. 5.7(d)), and list the general and special equivalent positions. Later we shall consider other methods for locating the position of a symmetry element that is equivalent to the combination of two others in a space group.

There are two sets of special positions, on 2, and the pairs of coordinates must be chosen correctly, so as to conform to *p2gg*. The selection may seem self-evident, but if not, the correct sets can be made up by inserting the coordinates of a special position into the coordinates of the general positions. Thus, for the two-fold rotation

† See footnote to page 161

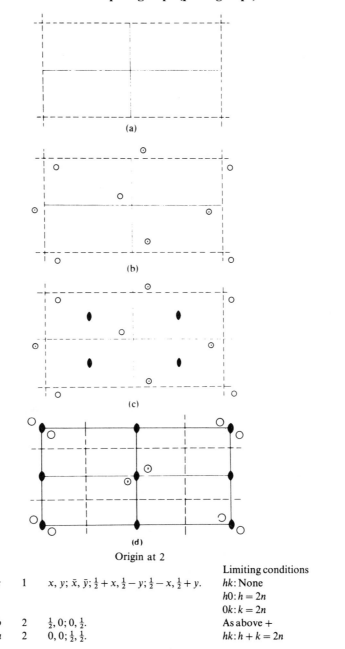

Origin at 2

					Limiting conditions
4	c	1	$x, y; \bar{x}, \bar{y}; \frac{1}{2}+x, \frac{1}{2}-y; \frac{1}{2}-x, \frac{1}{2}+y.$		hk: None
					$h0: h = 2n$
					$0k: k = 2n$
2	b	2	$\frac{1}{2}, 0; 0, \frac{1}{2}.$		As above +
2	a	2	$0, 0; \frac{1}{2}, \frac{1}{2}.$		$hk: h + k = 2n$

Fig. 5.7—Plane group $p2gg$: (a) unit cell framework of glide lines; origin at the intersection of gg, (b) general equivalent positions added, (c) additional symmetry elements included, (d) diagram reoriented with the origin on 2 (conventional), with description.

at the origin, $x = y = 0$, and we obtain the pair $0, 0$ and $\frac{1}{2}, \frac{1}{2}$. What plane group would have been obtained if we had paired $0, 0$ with $0, \frac{1}{2}$?

There are no special positions on glide lines, or on any translational symmetry element. Special positions, as sites for finite bodies, must conform to *point-group* symmetry. For this reason, plane group *pg* has no special positions at all.

5.3.3 Plane groups in the square system

We will consider one plane group in this system, because some new features arise with rotational symmetries of degree greater than 2. Consider *p4mm*. Since *4mm* is a point group, the origin is at *4mm*. Table 3.1 tells us again how the symmetry elements in the symbol are to be oriented. The space group diagram is readily completed by inserting a point x, y, operating on it by the symmetry *4mm*, and then repeating the *4mm* motif at the other three corners of the unit cell (Fig. 5.8).

If we now ask how we reach any point in the diagram from x, y by a single operation, we shall realize the existence of the four-fold rotation point at $\frac{1}{2}$, $\frac{1}{2}$, two-fold rotation points at 0, $\frac{1}{2}$ and $\frac{1}{2}$, 0, and diagonal glide lines as shown. The glide operation is envisaged as before: a reflexion across the line coupled with a translation of one-half the repeat distance in the direction of the line, that is, $a\sqrt{2}/2$ in this case.

The glide lines arise because we have a centred arrangement. If we look at Fig. 5.8 with the diagonal direction [11] as a side of a unit cell, that unit cell would have the plane group *c4mm*. This plane group is not separately recognized because it is the same as *p4mm*, and the rules for choice of unit cell (Section 4.3.1) dictate the use of the smaller cell.

The sets of coordinates are listed in the usual manner. The one new feature here, perhaps, is that by a 90° right-handed rotation the point x, y is transformed to $\bar{y}$, x. It may be appreciated readily from the geometry of the figure, and can be further justified through Appendix 5 ((A5.4), $\phi = 90°$).

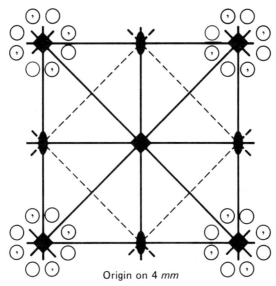

Origin on 4 *mm*

Limiting conditions

8	g	1	$x, y; \bar{x}, \bar{y}; y, x; \bar{x}, \bar{y}; x, \bar{x}, y; x, \bar{y}; x, \bar{y}, \bar{x}; y, x$	
4	f	m	$x, x; \bar{x}, \bar{x}; \bar{x}, x; x, \bar{x}$	
4	e	m	$x, \frac{1}{2}; \bar{x}, \frac{1}{2}; \frac{1}{2}, x; \frac{1}{2}, \bar{x}$	
4	d	m	$x, 0; \bar{x}, 0; 0, x; 0, \bar{x}$	
2	c	$2mm$	$\frac{1}{2}, 0; 0, \frac{1}{2}$	$hk: h + k = 2n$
1	b	$4mm$	$\frac{1}{2}, \frac{1}{2}$	
1	a	$4mm$	$0, 0$	

Fig. 5.8—Diagram and description of *p4mm*.

5.3.4 The seventeen plane groups

The seventeen plane groups are listed in Table 5.1 The reader should, after a consideration of the Problems, be able to deal satisfactorily with any of them. Fig. 5.9 illustrates the plane groups: there are two diagrams for each group, one to show the symmetry elements, and the other to show the general equivalent positions, using a scalene triangle instead of a circle to represent each position.

Table 5.1—The seventeen plane groups

System	Plane groups
Oblique	*p1, p2*
Rectangular	*pm, cm, pg*
	p2mm, c2mm, p2mg, p2gg
Square	*p4, p4mm, p4gm*
Hexagonal	*p3, p3m1, p31m*
	p6, p6mm

5.3.5 Comments on notation

Plane group *pm* may be written more fully as *pm1*, or even *p1m1*, so as to emphasize the fact that *m* is normal to the *x* axis (see Table 3.1). This notation is a trifle pedantic, so that *pm* will be used to indicate the standard setting of this group. The symbol *p1m* will mean that the *m* line is normal to the *y* axis. In the rectangular system, *pmm* implies *p2mm*. Nevertheless, the three symbols will be used and, in general, only the trival 1 operator will be omitted. An exception to this rule arises in the hexagonal system. While point-group symbols 3*m*1 and 31*m* are equivalent, corresponding only to a clockwise rotation of the *x*, *y* and *u* axes by 30° about the *z* axis, plane groups *p3m1* and *p31m* correspond to different arrangements of points in space, and so are not identical (see Fig. 5.9). The symmetry symbols in the two-dimensional space groups are collected together in Table 5.2.

Table 5.2—Symmetry symbols in two dimensions

Operation	Symbol	Graphic symbol
One-fold rotation about a point	1	None
Two-fold rotation about a point	2	⬤
Three-fold rotation about a point	3	▲
Four-fold rotation about a point	4	◆
Six-fold rotation about a point	6	⬢
Mirror reflexion across a line	*m*	————
Glide reflexion across a line	*g*	------

5.4 THREE-DIMENSIONAL SPACE GROUPS

We next extend the ideas arising from the previous sections into the third dimension, which will normally be implied by the term space group alone. There are 230 space groups, but we shall study a small number of them. We shall look mainly at the space groups in the low symmetry systems, and with good reason, because about 90% of the known crystalline substances crystallize in the triclinic, monoclinic and orthorhombic crystal systems.

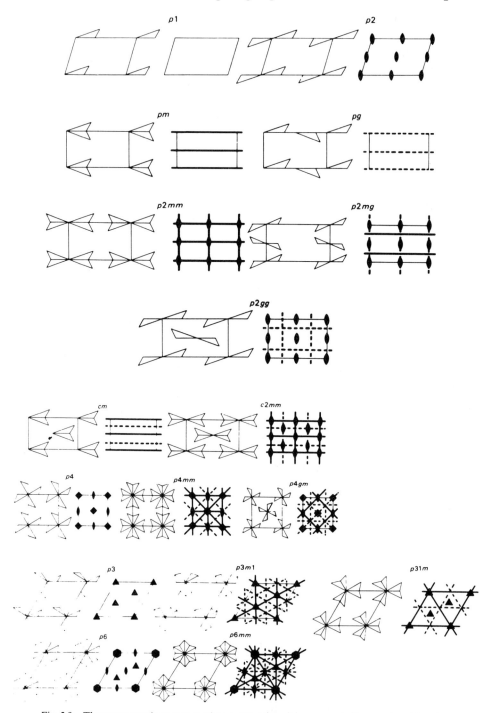

Fig. 5.9—The seventeen plane groups; the asymmetric unit is represented by a scalene triangle.

5.4.1 Triclinic space groups

There are two triclinic point groups, 1 and $\bar{1}$, and one triclinic lattice, P. We might anticipate space groups $P1$ and $P\bar{1}$, and they are the only triclinic space groups; $P\bar{1}$ is illustrated in Fig. 5.10.

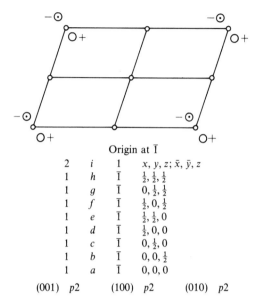

Origin at $\bar{1}$

2	i	1	$x, y, z; \bar{x}, \bar{y}, z$
1	h	$\bar{1}$	$\frac{1}{2}, \frac{1}{2}, \frac{1}{2}$
1	g	$\bar{1}$	$0, \frac{1}{2}, \frac{1}{2}$
1	f	$\bar{1}$	$\frac{1}{2}, 0, \frac{1}{2}$
1	e	$\bar{1}$	$\frac{1}{2}, \frac{1}{2}, 0$
1	d	$\bar{1}$	$\frac{1}{2}, 0, 0$
1	c	$\bar{1}$	$0, \frac{1}{2}, 0$
1	b	$\bar{1}$	$0, 0, \frac{1}{2}$
1	a	$\bar{1}$	$0, 0, 0$

(001) $p2$ (100) $p2$ (010) $p2$

Fig. 5.10—Diagram and description for triclinic space group $P\bar{1}$.

It is usual to choose the origin of a centrosymmetric space group on $\bar{1}$. In this space group, there are two general equivalent positions, and eight sets of special equivalent positions. We must remember that the space group diagrams are shown in projection, the $+$ and $-$ signs attached to each representative point indicating a z coordinate. The standard projection has $+x$ running from top to bottom, $+y$ running from left to right and $+z$ coming upwards from the plane of the diagram. As well as the four centres of symmetry shown in the diagram, at $0, 0, 0$, $0, \frac{1}{2}, 0$, $\frac{1}{2}, 0, 0$ and $\frac{1}{2}, \frac{1}{2}, 0$, we must note the other four centres on the plane $z = 1/2$ that are not shown explicitly. The point x, y, z, for example, is related to the point $\bar{x}, \bar{y}, 1 - z$ by a centre of symmetry at $0, 0, \frac{1}{2}$. This fact may be checked by drawing the space group in the ac projection. Sometimes we may need to check for special positions in this way; after longer study, we shall come to know what to expect.

The principal projections of a three-dimensional space group will correspond to plane group symmetries. In $P\bar{1}$, each projection has the plane group $p2$; 2 is the two-dimensional analogue of $\bar{1}$.

5.4.2 Monoclinic space groups

In the monoclinic system we have point groups 2, m and $2/m$ and two lattices, P and C, to consider. Here and above (5.4.1), we have slipped into the jargon of 'P lattice', and etc—but note Section 4.4.

Space groups related to point group 2

Just as we obtained *p*2 from a two-fold symmetry motif and the oblique lattice, so we can obtain the space group *P*2 from a similar motif and a monoclinic *P* lattice (Fig. 5.11). The origin is on 2, and is defined, therefore, only with respect to *x* and *z* until the first general position *x*, *y*, *z* is chosen. The graphic symbol for 2 in the plane of the diagram is ⟶.

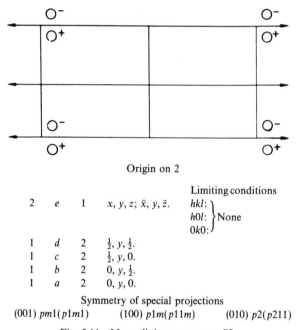

Origin on 2

				Limiting conditions
2	*e*	1	$x, y, z; \bar{x}, y, \bar{z}.$	$hkl:$
				$h0l:$ }None
				$0k0:$
1	*d*	2	$\frac{1}{2}, y, \frac{1}{2}.$	
1	*c*	2	$\frac{1}{2}, y, 0.$	
1	*b*	2	$0, y, \frac{1}{2}.$	
1	*a*	2	$0, y, 0.$	

Symmetry of special projections

(001) *pm*1(*p*1*m*1)   (100) *p*1*m*(*p*11*m*)   (010) *p*2(*p*211)

Fig. 5.11—Monoclinic space group *P*2.

The general and special equivalent positions are derived quite readily; the sets *b* and *d* refer to diad axes at the level *c*/2 (*z* = 1/2).

Fig. 5.12 illustrates space group *C*2. It may be considered either as the two-fold symmetry motif coupled to a monoclinic *C* lattice, or as *P*2 with the centring condition $\frac{1}{2}, \frac{1}{2}, 0$ applied to both the symmetry elements and the equivalent positions; the parenthetical expression (0, 0, 0; $\frac{1}{2}, \frac{1}{2}, 0$)+ in Fig. 5.12 refers to the latter procedure. Fig. 5.15 is a stereoview of the unit cell and general equivalent positions in *C*2.

In *C*2 there are only two sets of special positions compared with four sets in *P*2. The reason for this situation is similar to that discussed for *pm* and *cm* (Section 5.3.2).

Two new features arise from the study of *C*2. When considering the projection of the space group on to the principal planes containing one of the centring components, that is, the *ac* or the *bc* plane, we find two translation repeats in projection. Thus, it is sufficient to refer the plane group of the projection to a smaller repeat unit, for example, on (100) the repeat *b*′ is equal to *b*/2.

A simple way of obtaining the plane group of the projection, where it cannot be done by inspection, is to plot the projected coordinates, making sure that the shape of the projected unit cell is chosen correctly. Thus, for the (100) projection of *C*2, we plot y, z; $y, \bar{z}$; $\frac{1}{2} + y, \frac{1}{2} + z$ and $\frac{1}{2} + y, \frac{1}{2} - z$ on a rectangular unit cell. Then, we require the plane group symmetry that would relate the points on the projection. With

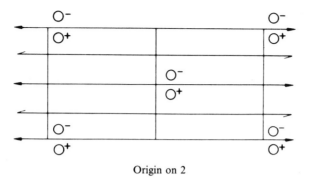

Origin on 2

			$(0, 0, 0; \frac{1}{2}, \frac{1}{2}, 0) +$	Limiting conditions
4	c	1	$x, y, z; \bar{x}, y, \bar{z}.$	$hkl: h + k = 2n$ $h0l: (h = 2n)$ $0k0: (k = 2n)$
2	b	2	$0, y, \frac{1}{2}.$	As above
2	a	2	$0, y, 0.$	

Symmetry of special projections

(001) $cm1(c1m1)$ (100) $p1m(p11m)$ $b' = b/2$ (010)$p2(p211)$ $a' = a/2$

Fig. 5.12—Monoclinic space group $C2$.

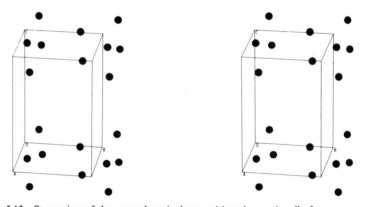

Fig. 5.13—Stereoview of the general equivalent positions in a unit cell of space group $C2$.

right-handed axes $x'(y)$ and $y'(z)$, the m line is normal to the new, y' axis, and we arrive at the symbol $p1m$ ($p11m$), with $b' = b/2$ (Fig. 5.14).

A question that is often asked is how points of the same handedness, for example, x, y, z and $\bar{x}, y, \bar{z}$, as in $P2$, become of opposite handedness, y, z and $y, \bar{z}$, in projection. It is suggested that the reader make a drawing of $P2$ using ♀ and ♀ rather than $\bigcirc^+$ and $\bigcirc^-$, and then remember that the plane group is strictly two-dimensional (see also Section 3.6.6 and Fig. 3.23).

The second new feature to emerge from the study of $C2$ is the screw axis. Just as a centring condition in conjunction with a reflexion symmetry gives rise to glide symmetry (Section 5.3.2), so centring in conjunction with rotational symmetry (in three dimensions) give rise to screw axis symmetry. If, in considering $C2$, we have

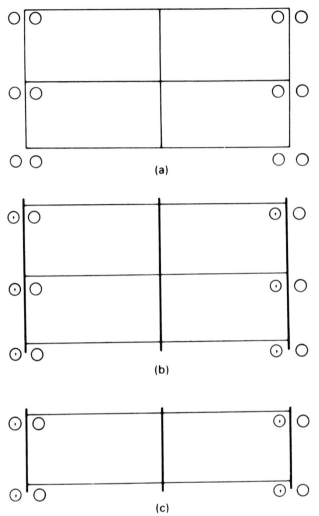

Fig. 5.14—Projection of $C2$ on to (100): (a) the y, z positions from $C2$ (z axis to right), (b) two-dimensional symmetry elements added, (c) one unit cell of plane group $p1m$ ($p11m$), $b' = b/2$, $c' = c$.

asked the question 'what single symmetry operation relates x, y, z to $\frac{1}{2} - x, \frac{1}{2} + y, \bar{z}$', we shall have realized the need for a symmetry operation other than 2 or C. It is, in fact, the 2_1 screw axis, graphic symbol   in the plane of projection.

5.4.3 Screw axes

Screw axis symmetry arises in three-dimensional space groups of monoclinic and higher symmetry. A screw axis is a single symmetry element, but couples both rotational and translational movements. Consider first the more familiar spiral staircase (Fig. 5.15). Imagine that the bottom stair is rotated about the vertical support (axis) by an anticlockwise rotation of 60°, coupled with a translation in the direction of the axis by one-sixth of the distance between similarly oriented stairs, so

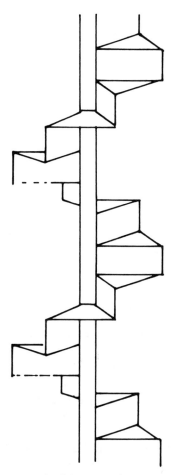

Fig. 5.15—Spiral staircase: an example of 6_1 screw axis symmetry; for clockwise rotation, the axis would be designated 6_5.

that it takes the place of the second stair, the second stair moving into the place of the third stair, and so on. After six such operatons, we arrive at a stair in the same orientation as the bottom step, but one repeat distance above it. Clearly, if the staircase were of infinite length (the 'Monument' or the 'Statue of Liberty'?), each operation would cause the staircase to be mapped on to itself. We call the symmetry operation here a 6_1 screw axis.

In general, a screw axis is of the type R_n ($R = 2, 3, 4, 6; n < R$), and the operation consists of an R-fold rotation coupled with a translation of n/R times the repeat distance in the direction of the axis (if $n = R$, $R_n \equiv R$). In $C2$, the centring introduces the screw diads which interleave the diads as shown in Fig. 5.12; their translational component is $b/2$.

From our discussion of pm and cm, it should follow that $C2_1$ is equivalent to $C2$, but that $P2_1$ is a different, new space group (Fig. 5.16). We may note that there are no special positions in $P2_1$ because this space group contains only translational symmetry elements. Furthermore, whereas a two-fold rotation axis projects parallel to the axis as an m line, a two-fold screw axis projects parallel to itself as a g line.

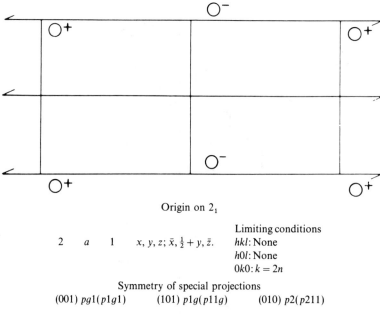

Origin on 2_1

$$2 \quad a \quad 1 \quad x, y, z; \bar{x}, \tfrac{1}{2} + y, \bar{z}.$$

Limiting conditions
hkl: None
$h0l$: None
$0k0: k = 2n$

Symmetry of special projections
(001) $pg1(p1g1)$ (101) $p1g(p11g)$ (010) $p2(p211)$

Fig. 5.16—Space group $P2_1$.

Space groups related to point group m

If we carry out constructions like those for pm and cm (Section 5.3.2) in relation to the monoclinic lattices, we shall obtain space groups Pm and Cm. The centring in cm introduced glide lines; that in Cm introduces glide *planes*. This feature is new to our discussion, whereas the other aspects of Pm and Cm are similar to pm and cm but with the geometrical extension to the third dimension. We will discuss glide planes next, and continue the study of the monoclinic space groups in that section.

5.4.4 Glide planes

Let Fig. 5.2(b) be regarded as a drawing of a three-dimensional space group, and let the dashed lines be the traces of a glide plane normal to b. In the two-dimensional glide line, the direction of the glide movement after reflexion is unequivocal. In three dimensions, several possibilities arise. In the case of the glide plane normal to b, the translational movement could along a (an a glide plane, with a component $a/2$), along c (a c glide plane, with a component of $c/2$), along a diagonal direction, $\mathbf{a} + \mathbf{c}$ (an n glide plane with a component $(a + c)/2$), along the diagonal direction (a d glide, with a component of $(a \pm c)/4$)—so named because it is found in the space group of diamond—but not, of course, along b, the direction of the reflexion. The different possibilities do not necessarily give rise to differing space groups. Let us write down the symbols that might seem to arise:

$Pm*$ $Cm*$

Pa Ca

$Pc*$ $Cc*$

Pn Cn

Only those symbols marked with an * are distinct. We may be able to see straightaway the equivalence between Pa and Pc and between Cm and Ca, but we shall investigate these space groups more fully through the Problems section.

Space groups related to point group 2/m
By comparison with the monoclinic space groups already discussed, we can write down

$$P2/m \xrightarrow{\quad +C \quad} C2/m$$
$$P2_1/m \xrightarrow{\quad +C \quad}$$
$$P2/c \xrightarrow{\quad +C \quad} C2/c$$
$$P2_1/c \xrightarrow{\quad +C \quad}$$

avoiding immediately any equivalent, unnecessary symbols. As an example, we shall study $P2_1/c$, a space group that is encountered frequently in practice.

This space group, related to point group $2/m$, must be centrosymmetric; the centre of symmetry, to be chosen as an origin, does not lie at the intersection of 2_1 and the c glide plane. We could approach this space group in the manner of $p2gg$. However, here we shall adopt an analytical approach based on the full meaning of the symbol. From Table 3.5, and our choice of origin, we can write the following specification:

$$\bar{1} \text{ at } 0, 0, 0$$

$$2_1 \text{ along } [p, y, r]\dagger$$

$$c \text{ is the plane } (x, q, z)$$

where p, q and r are to be determined. We carry out these symmetry operations on a point x, y, z according to the following scheme:

$$(1)\ x, y, z \xrightarrow{\qquad\qquad c \qquad\qquad} (2) \qquad x, 2q - y, \tfrac{1}{2} + z$$

$$\downarrow 2_1$$

$$(3)\ 2p - x, \tfrac{1}{2} + 2q - y, 2r - \tfrac{1}{2} - z$$

$$\xrightarrow{\quad \bar{1} \quad} (4) \qquad \bar{x} \qquad \bar{y} \qquad \bar{z}$$

We now use the fact that the combination of two symmetry operations is equivalent here to a third operation, performed on the original point (1). Symbolically, $2_1 \cdot c \equiv \bar{1}$, and points (3) and (4) are one and the same. Hence, by comparing coordinates, $p = 0$ and $q = r = 1/4$. (Strictly, $q = -1/4$, but 1/4 is crystallographically equivalent.) These results lead to the desired diagram for $P2_1/c$ (Fig. 5.17).

The change in the y coordinate by the c operation (1) to (2) is illustrated in Fig. 5.18. A similar argument can be applied to any corresponding situation. We shall complete the details of $P2_1/c$ by means of a Problem.

The centred monoclinic space groups of this class will not be discussed here; they present little difficulty once the primitive space groups have been mastered. Fig. 5.19 shows stereoviews of the molecule and unit cell of diiodo-$(N,N,N',N'$-tetramethylethylenediaine)zinc(II), $I_2[(CH_3)_2NCH_2CH_2N(CH_3)_2]Zn$, which crystallizes in space group $C2/c$. The unit cell dimensions are $a = 1.312$, $b = 0.783$, $c = 1.357$ nm, $\beta = 111.4°$,

---

† This notation means the line parallel to y and at $x = p$, $z = r$.

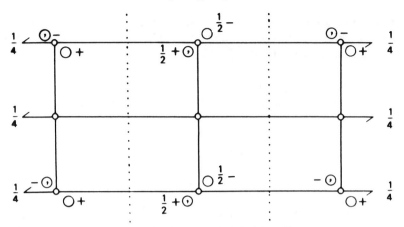

Fig. 5.17—Space group $P2_1/c$ (origin at $\bar{1}$).

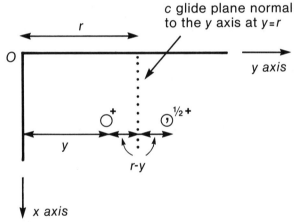

Fig. 5.18—Operation of a c glide with the plane normal to the y axis and cutting it at distance r from the origin.

and the density D_m is 2202 kg m^{-3}. The reader is invited to determine the number of molecules per unit cell, using (5.3), set up a diagram for $C2/c$, and consider what can be determined at this stage about the positions of the molecules in the unit cell:

$$D_m = ZM_r u/V \qquad (5.3)$$

where M_r is the relative molecular mass, u the atomic mass unit (1.6606 × 10^{-27} kg) and V is the volume of the unit cell; Z must be an integer, and any discrepancy will be a reflection of experimental errors in the other parameters of (5.3), principally in D_m and V.

5.4.5 Summary of monoclinic space groups
The monoclinic space groups, thirteen in number, are listed in Table 5.3. From time to time, crystal structures are listed in non-standard space groups, for example, $I2$ or $P2_1/n$. We must recognize such symbols, and their equivalent notations in Table 5.3.

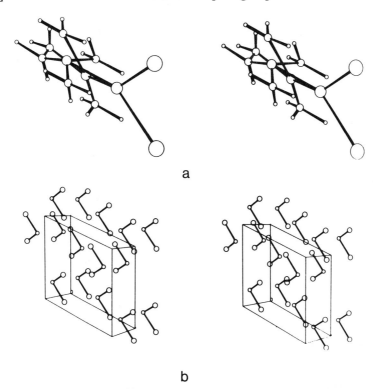

a

b

Fig. 5.19—Stereoviews of the structure of diiodo-(N,N,N',N'-tetramethylethylenediamine)zinc(II): (a) structure; the circles, in decreasing order of size, represent I, Zn, N, C and H, (b) unit cell, showing only the Zn and I atoms; the two Zn–I bonds are of equal length.

Table 5.3—Monoclinic space groups•

$P2$	$C2$	Pm	Cm	$P2/m$	$C2/m$
$P2_1$		Pc	Cc	$P2_1/m$	
				$P2/c$	$C2/c$
				$P2_1/c$	

5.4.6 Half-shift rule

In the space groups belonging to the $2/m$ class, we have four possible situations to consider for the origin (Fig. 5.20). Only in $2/m$ does the symmetry axis intersect the reflexion plane in the centre of symmetry (a). In the other three cases, this intersection, marked * in (b), (c) and (d), is translated from the position of $\bar{1}$, marked ○, by virtue of the translational symmetry elements 2_1, c or both 2_1 and c. From Fig. 5.18 we saw that a shift of a symmetry element by an amount r caused a shift in a coordinate by an amount $2r$ from the position it would have occupied had the symmetry element passed through the point $y = 0$. Here, we need to work in the opposite manner, as

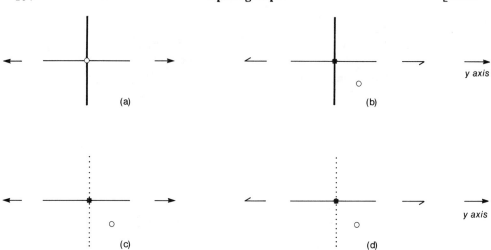

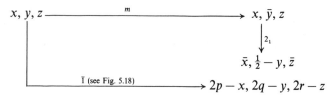

Fig. 5.20—Half-shift rule: (a) $2/m$, (b) $2_1/m$, (c) $2/c$, (d) $2_1/c$; the * marks the intersection of the symmetry axis and the symmetry plane, and O represents the centre of symmetry.

follows. Let m be the plane $(x, 0, z)$, 2_1 the line $[0, y, 0]$ and $\bar{1}$ the point p, q, r. Then, we have

Hence, $\bar{1}$ is at $0, \frac{1}{4}, 0$. In other words, to obtain the position of the centre of symmetry with respect to the point of intersection of the symmetry axis and the symmetry plane, we take the total amount of translation shift and halve it, to determine the amount by which the point * is shifted from $\bar{1}$ at $0, 0, 0$. In $P2_1/m$ it is $b/4$, that is, the m plane is moved to $(x, \frac{1}{4}, z)$, with 2_1 remaining at $[0, y, 0]$. Applying this rule to $P2_1/c$, we have a total shift of $(b/2 + c/2)$, and the half shift is $(b/4 + c/4)$. Hence, for $\bar{1}$ at the origin, we must position c at $(x, \frac{1}{4}, z)$ and 2_1 at $[0, y, \frac{1}{4}]$, just as determined earlier (Fig. 5.17). We shall find this rule to be of great use in the monoclinic and orthorhombic space groups.

5.4.7 Orthorhombic space groups

The orthorhombic space groups are a little more complex. We have point groups 222, $mm2$ and mmm, together with four lattices, P, C, I and F. In addition, we find that although the *lattice* descriptors C and A are equivalent, indicating only a change of axes, in combination with $mm2$ they lead to the differing space groups $Cmm2$ and $Amm2$. Thus, we must consider also the A lattice when studying space groups related to point group $mm2$.

Space groups related to point group 222

We may begin by combining point group 222 with the four lattices:

$$P222 \qquad C222 \qquad I222 \qquad F222$$
$$\text{(a)} \qquad \text{(b)} \qquad \text{(c)} \qquad \text{(d)}$$

With type (a) we may introduce a 2_1 axis, leading to $P222_1$. Since there is no restriction, other than convention, on the naming of axes, $P22_12$ and $P2_122$ are equivalent to the standard setting $P222_1$. A second 2_1 axis leads to $P2_12_12$, equivalent to $P2_122_1$ and $P22_12_1$.

The introduction of a third two-fold screw axis raises an interesting feature. We recall from Euler's construction (Section 3.6.4) that a third axis in a symbol is not independent of the other two axes. If any two perpendicular 2_1 axes intersect, then space group $P2_12_12$, or an equivalent, is obtained. In $P2_12_12_1$ the three mutually perpendicular 2_1 axes *do not* intersect one another. We shall continue our study of this space group later.

In type (b), we recall that C centring introduces translations of $\frac{1}{2}$ along a and b, and so leads to 2_1 axes interleaving the two-fold axes parallel to the x and y axes. Hence, $C2_122$, $C22_12$ and $C2_12_12$ are all equivalent to $C222$. But $C222_1$ is new because the translation of $c/2$ imposed by the presence of the 2_1 axis is not already provided by the C centring; $C222_1$ is, however, identical to $C2_12_12_1$. It follows that $F222$ is the only space group of type (d).

In space group $I222$ there are three intersecting diads and three intersecting screw diads. If we couple the body-centring condition to $P2_12_12_1$, we obtain a space group with three non-intersecting diads and three non-intersecting screw diads. By convention, the latter space group is $I2_12_12_1$. We have here two space groups, $I222$ and $I2_12_12_1$ there are not uniquely specified by their Hermann–Mauguin notation; the conventional choice has been given above.

We may summarise these space groups by the following scheme:

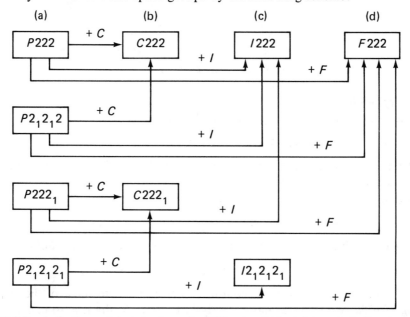

In $P2_12_12$ only the 2_1 axes intersect, whereas in $P222_1$ a two-fold axis intersects the 2_1 axis.

Space group $P2_12_12_1$ is of relatively common occurrence. There is more than one way of setting three mutually perpendicular non-intersecting axes, but the standard setting is shown in Fig. 5.21. Although $P2_12_12_1$ is a non-centrosymmetric space group, each of the three principal projections has the plane group $p2gg$; thus, they are centrosymmetric and are set up in accordance with Fig. 5.7(d). This setting, with respect to $P2_12_12_1$, involves a change of origin from that given in Fig. 5.21.

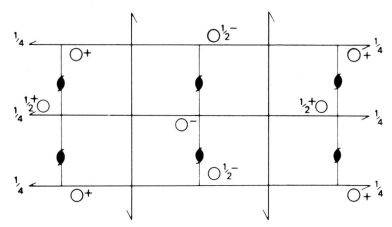

Origin halfway between three pairs of non-intersecting screw axes

<table>
<tr><td></td><td></td><td></td><td></td><td colspan="2">Limiting conditions</td></tr>
<tr><td>4</td><td>a</td><td>1</td><td>$x, y, z; \frac{1}{2} - x, \bar{y}, \frac{1}{2} + z; \frac{1}{2} + x, \frac{1}{2} - y, \bar{z}; \bar{x}, \frac{1}{2} + y, \frac{1}{2} - z.$</td><td>$hkl:$</td><td></td></tr>
<tr><td></td><td></td><td></td><td></td><td>$0kl:$</td><td rowspan="2">None</td></tr>
<tr><td></td><td></td><td></td><td></td><td>$h0l:$</td><td></td></tr>
<tr><td></td><td></td><td></td><td></td><td>$hk0:$</td><td></td></tr>
<tr><td></td><td></td><td></td><td></td><td>$h00: h = 2n$</td><td></td></tr>
<tr><td></td><td></td><td></td><td></td><td>$0k0: k = 2n$</td><td></td></tr>
<tr><td></td><td></td><td></td><td></td><td>$00l: l = 2n$</td><td></td></tr>
</table>

Symmetry of special projections
(001) $p2gg$ (100) $p2gg$ (010) $p2gg$

Fig. 5.21—Space group $P2_12_12_1$.

Change of origin
Consider the projection of $P2_12_12_1$ on to (001); the general equivalent positions are:

$$x, y; \quad \tfrac{1}{2} - x, \bar{y}; \quad \tfrac{1}{2} + x, \tfrac{1}{2} - y; \quad \bar{x}, \tfrac{1}{2} + y$$

Let the new origin be at $\frac{1}{4}, 0$ in the diagram of $P2_12_12_1$: a 2_1 axis projects down its length as a two-fold rotation point. Then, the coordinates of the new origin are subtracted from the original coordinates:

$$x - \tfrac{1}{4}, y; \quad \tfrac{1}{4} - x, \bar{y}; \quad \tfrac{1}{4} + x, \tfrac{1}{2} - y; \quad -x - \tfrac{1}{4}, \tfrac{1}{2} + y$$

Next, new variables are assigned such that $x - \frac{1}{4} = x_0$ and $y = y_0$, leading to

$$x_0, y_0; \quad \bar{x}_0, \bar{y}_0; \quad \tfrac{1}{2} + x_0, \tfrac{1}{2} - y_0; \quad \tfrac{1}{2} - x_0, \tfrac{1}{2} + y_0$$

and, without the subscript, these coordinates correspond to those of the standard setting of $p2gg$. We shall see more clearly in Chapter 6 why it is so convenient to choose a space group origin on $\bar{1}$ (three dimensions) or 2 (two dimensions).

Space groups related to point group mm2

In the space groups related to this crystal class, we have two symmetry planes (reflexion and/or glide) intersecting in a line which is a symmetry axis (2 or 2_1) or is parallel to such an axis. The space group origin is chosen on 2 (or 2_1), that is, the axis is the line $[0, 0, z]$. In space group $Pmn2_1$, it is usual for the origin to be taken at the intersection of the m and n symmetry planes: it leads to a simpler expression of the coordinates of the general equivalent positions, although it is not incorrect to take the origin on 2_1. In order to illustrate some new space group features, we shall consider $Pma2$, $Pma2 + C$ and $Pma2 + A$.

From Table 3.5, we have that in $Pma2$, the m plane is normal to the x axis, and the a glide plane is normal to the y axis. Applying the half-shift rule, we see that the line of intersection of m and a is set off from $[0, 0, z]$ by $a/2$; thus, we obtain Fig. 5.22.

The effect of C centring on $Pma2$ is easily found by applying the translations $\frac{1}{2}, \frac{1}{2}, 0$ to the general equivalent positions in $Pma2$. The completion of the diagram

$$P m a 2$$
$$C_{2v}^4$$

No. 28 $P\,m\,a\,2$ $m\,m\,2$ Orthorhombic

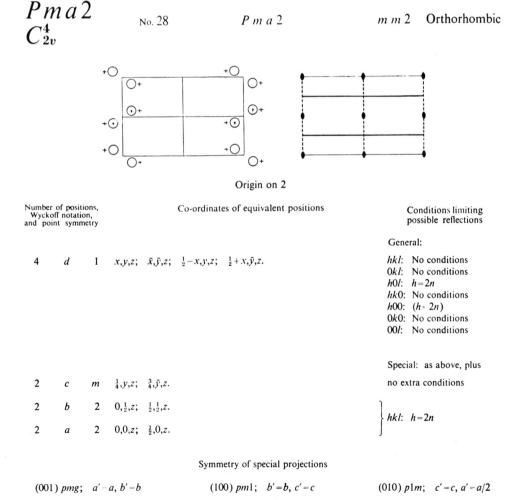

Origin on 2

Number of positions, Wyckoff notation, and point symmetry			Co-ordinates of equivalent positions	Conditions limiting possible reflections

General:

| 4 | d | 1 | $x,y,z;\quad \bar{x},\bar{y},z;\quad \frac{1}{2}-x,y,z;\quad \frac{1}{2}+x,\bar{y},z.$ | |

hkl: No conditions
$0kl$: No conditions
$h0l$: $h=2n$
$hk0$: No conditions
$h00$: $(h=2n)$
$0k0$: No conditions
$00l$: No conditions

Special: as above, plus

no extra conditions

2	c	m	$\frac{1}{4},y,z;\quad \frac{3}{4},\bar{y},z.$	
2	b	2	$0,\frac{1}{2},z;\quad \frac{1}{2},\frac{1}{2},z.$	
2	a	2	$0,0,z;\quad \frac{1}{2},0,z.$	

$\left.\begin{array}{c}\\ \\\end{array}\right\}$ hkl: $h=2n$

Symmetry of special projections

(001) pmg; $a' = a$, $b' = b$ (100) $pm1$; $b' = b$, $c' = c$ (010) $p1m$; $c' = c$, $a' = a/2$

Fig. 5.22—Space group $Pma2$.

Orthorhombic $m\,m\,2$ $C\,m\,m\,2$ No. 35

$$C\,m\,m\,2$$
$$C_{2v}^{11}$$

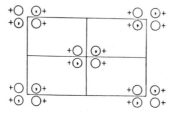 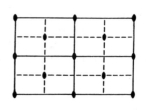

Origin on $mm2$

Number of positions, Wyckoff notation, and point symmetry	Co-ordinates of equivalent positions $(0,0,0;\ \tfrac{1}{2},\tfrac{1}{2},0)+$	Conditions limiting possible reflections

General:

8   f 1 $x,y,z;\ \ \bar{x},\bar{y},z;\ \ \bar{x},y,z;\ \ x,\bar{y},z.$

                                                hkl: $h+k=2n$
                                                $0kl$: $(k=2n)$
                                                $h0l$: $(h=2n)$
                                                $hk0$: $(h+k=2n)$
                                                $h00$: $(h=2n)$
                                                $0k0$: $(k=2n)$
                                                $00l$:   No conditions

Special: as above, plus

4   e m $0,y,z;\ \ 0,\bar{y},z.$

4   d m $x,0,z;\ \ \bar{x},0,z.$ } no extra conditions

4   c 2 $\tfrac{1}{4},\tfrac{1}{4},z;\ \ \tfrac{1}{4},\tfrac{3}{4},z.$ hkl: $h=2n;$ $(k=2n)$

2   b mm $0,\tfrac{1}{2},z.$

2   a mm $0,0,z.$ } no extra conditions

Symmetry of special projections

$(001)\,cmm;$ $a'=a,\ b'=b$ $(100)\,pm1;$ $b'=b/2,\ c'=c$ $(010)\,p1m;$ $c'=c,\ a'=a/2$

Fig. 5.23—Space group $Cmm2$ ($Cma2$).

reveals further m planes at $(x,\tfrac{1}{4},z)$ and $(x,\tfrac{3}{4},z)$, b glide planes at $(0,y,z)$ and $(\tfrac{1}{2},y,z)$ and diads at $[\tfrac{1}{4},\tfrac{1}{4},z]$, $[\tfrac{3}{4},\tfrac{1}{4},z]$, $[\tfrac{1}{4},\tfrac{3}{4},z]$ and $[\tfrac{3}{4},\tfrac{3}{4},z]$. The conventional symbol for this space group is $Cmm2$, with the origin on $mm2$. It could be obtained also from $Pmm2+C$, for which we should expect to find the glide planes interleaving both sets of m planes (Fig. 5.23).

$Pma2+A$ is shown in Fig. 5.24. The A centring introduces translations of $b/2$ and $c/2$ into the coordinate sets, which leads to n glide planes at $y=\tfrac{1}{4}$ and $\tfrac{3}{4}$, interleaving the a glide planes, and at $x=\tfrac{1}{4}$ and $\tfrac{3}{4}$, coincident with the m planes; there are also 2_1 axes interleaving the two-fold axes, at $\pm[0,\tfrac{1}{4},z]$ and $\pm[\tfrac{1}{2},\tfrac{1}{4},z]$.

These examples show that we can treat a centred group as the corresponding primitive space group plus the appropriate centring condition. We must seek all the symmetry elements present, according to the procedures already established. The

$A\,m\,a\,2$
C^{16}_{2v}

No. 40 $A\ m\ a\ 2$ $m\ m\ 2$ Orthorhombic

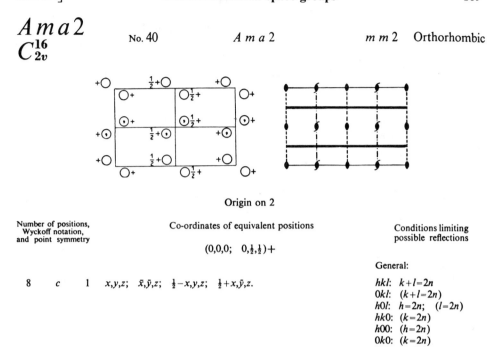

Origin on 2

Number of positions, Wyckoff notation, and point symmetry			Co-ordinates of equivalent positions $(0,0,0;\ 0,\tfrac{1}{2},\tfrac{1}{2})+$	Conditions limiting possible reflections

Co-ordinates of equivalent positions

$(0,0,0;\ 0,\tfrac{1}{2},\tfrac{1}{2})+$

Conditions limiting possible reflections

General:

8   c 1 $x,y,z;\ \bar{x},\bar{y},z;\ \tfrac{1}{2}-x,y,z;\ \tfrac{1}{2}+x,\bar{y},z.$

$hkl:\ k+l=2n$
$0kl:\ (k+l=2n)$
$h0l:\ h=2n;\ \ (l=2n)$
$hk0:\ (k=2n)$
$h00:\ (h=2n)$
$0k0:\ (k=2n)$
$00l:\ (l=2n)$

Special: as above, plus

4   b m $\tfrac{1}{4},y,z;\ \tfrac{3}{4},\bar{y},z.$

no extra conditions

4   a 2 $0,0,z;\ \tfrac{1}{2},0,z.$

$hkl:\ h=2n$

Symmetry of special projections

(001) $pmg;\ a'=a,\ b'=b/2$ (100) $cm1;\ b'=b,\ c'=c$ (010) $p1m;\ c'=c/2,\ a'=a/2$

Fig. 5.24—Space group $Ama2$.

standard space group symbol may usually be obtained by the rule of precedence in naming: where there is more than one type of symmetry element present, parallel to a given plane or line, then the precedence is $m>a>b>c>n>d$ and $2>2_1$. Hence, $Cma2$ becomes $Cmm2$, but $Ama2$ remains $Ama2$. All other space groups of this class can be treated by the methods herein described.

The third symbol, 2 or 2_1, need not be given before the space group can be set up. We can decide between 2 and 2_1 by the nature of the other translations present. Thus, in Pma there is no translation along the z axis; hence, we obtain $Pma2$. In Pca, the c glide translation leads to 2_1 symmetry along z, and we obtain $Pca2_1$. Where the symmetry planes provide two translations along z, they combine to produce pure rotational symmetry along z, as in $Pnn2$.

Space groups related to point group mmm

The centrosymmetric point group mmm has the full symbol $\dfrac{2}{m}\dfrac{2}{m}\dfrac{2}{m}$. There are several symmetry elements to consider and many combinations of them are permissible. The intersection of three mutually perpendicular m planes defines a centre of symmetry. When one or more of the symmetry planes is a glide plane, the centre of symmetry will be translated, according to the nature of the glide planes present. In order to obtain a standard setting of a space group in this class, we can use the half-shift rule. However, before tackling a space group in this manner, a little analysis may prove instructive.

The following scheme shows the sign changes for a point x, y, z under the symmetry operations of *mmm*:

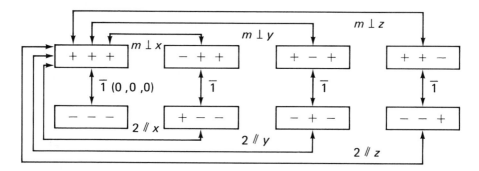

With the aid of this scheme, we can write down the coordinates for an orthorhombic space group, for example, *Pnma*. The space group symbol leads to the following specification:

$$n \quad (p, y, z) \qquad m \quad (x, q, z) \qquad a \quad (x, y, r)$$
$$2_P \quad [x, B, C] \qquad 2_Q \quad [A, y, C'] \qquad 2_R \quad [A', B', z]$$
$$\bar{1} \quad 0, 0, 0 \text{ (choice of origin)}$$

The eight general equivalent positions can now be given, as follows:

(1) x, y, z (2) $2p - x, \frac{1}{2} + y, \frac{1}{2} + z$ (3) $x, 2q - y, z$ (4) $\frac{1}{2} + x, y, 2r - z$

(5) $\bar{x}, \bar{y}, \bar{z}$ (6) $2p + x, \frac{1}{2} - y, \frac{1}{2} - z$ (7) $\bar{x}, 2q + y, \bar{z}$ (8) $\frac{1}{2} - x, \bar{y}, 2r + z$

Pairs of cordinates such as 1, 2 and 4, 7, or 1, 3 and 2, 8 are related by the same symmetry operations. Hence, it follows that $p = q = r = 1/4$. Furthermore, the pairs such as 1, 6 show that $P = 1$, $B = C = 1/4$, $Q = 1$, $A = C = 0$, and $R = 1$, $A' = 1/4$, $B' = 0$.

We now have the nature and orientation of all of the symmetry elements in *Pnma*, and the coordinates of the general equivalent positions. The full symbol of this centrosymmetric space group may be written $P\dfrac{2_1}{n}\dfrac{2_1}{m}\dfrac{2_1}{a}$ and it is illustrated by Fig. 5.25. The analysis has revealed clearly the useful facts, applicable to all triclinic, monoclinic and orthorhombic space groups (and to some others), that reflexion produces a single sign change in coordinates, the reflexion plane being normal to the coordinate direction that changes sign, that rotation produces two sign changes, those other than in the direction of the axis, and a centre of symmetry produces three sign changes. The other variations in the coordinates depend first upon whether or no we are dealing with translational symmetry elements, and secondly upon the positions of the symmetry elements with respect to the origin. Thus, if a coordinate

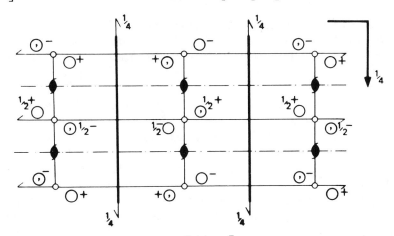

Origin at $\bar{1}$

8   d   1 $x, y, z; \frac{1}{2}+x, \frac{1}{2}-y, \frac{1}{2}-z; \bar{x}, \frac{1}{2}+y, \bar{z}; \frac{1}{2}-x, \bar{y}, \frac{1}{2}+z;$ Limiting conditions
            $\bar{x}, \bar{y}, \bar{z}; \frac{1}{2}-x, \frac{1}{2}+y, \frac{1}{2}+z; x, \frac{1}{2}-y, z; \frac{1}{2}+x, y, \frac{1}{2}-z.$ hkl: None
            $okl: k + l = 2n$
            $h0l$: None
            $hk0: h = 2n$
            $h00: (h = 2n)$
            $0k0: (k = 2n)$
            $00l: (l = 2n)$

4   c   m $x, \frac{1}{4}, z; \bar{x}, \frac{3}{4}, \bar{z}; \frac{1}{2}-x, \frac{1}{4}, \frac{1}{2}+z; \frac{1}{2}+x, \frac{3}{4}, \frac{1}{2}-z.$ As above

4   b $\bar{1}$ $0, 0, \frac{1}{2}; 0, \frac{1}{2}, \frac{1}{2}; \frac{1}{2}, 0, 0; \frac{1}{2}, \frac{1}{2}, 0.$ As above +

4   a $\bar{1}$ $0, 0, 0; 0, \frac{1}{2}, 0; \frac{1}{2}, 0, \frac{1}{2}; \frac{1}{2}, \frac{1}{2}, \frac{1}{2}.$ $hkl: h + l = 2n; k = 2n$

Symmetry of special projections
(001) $p2gm$ (100) $c2mm$ (010) $p2gg$

Fig. 5.25—Space group $Pnma$.

x is related through a symmetry element to $s - x$, then that symmetry element passes through the point $x = s/2$ (see Fig. 5.18).

We can approach the derivation of the space group information by a scheme similar to that used for $P2_1/c$, remembering that

$$m_z \cdot m_y \cdot m_x \equiv \bar{1} \tag{5.4}$$

or, in this example,

$$a \cdot m \cdot n \equiv \bar{1} \tag{5.5}$$

In the cumulative operations with this scheme, a point x, y, z is moved to $\bar{x}, \bar{y}, \bar{z}$ by means of three successive reflexion operations, subject to the sum of the translational components of each coordinate over the cycle of operations being zero, because $\bar{1}$ is at $0, 0, 0$:

$$\Sigma t_x = \Sigma t_y = \Sigma t_z = 0 \tag{5.6}$$

$$x, y, z \xrightarrow{\quad n \quad} 2p - x, \tfrac{1}{2} + y, \tfrac{1}{2} + z$$

$$\bar{1} \uparrow \qquad\qquad\qquad\qquad\qquad \downarrow m$$

$$\tfrac{1}{2} + 2p - x, 2q - \tfrac{1}{2} - y, 2r - \tfrac{1}{2} - z \xleftarrow{\quad a \quad} 2p - x, 2q - \tfrac{1}{2} - y, \tfrac{1}{2} + z$$

whence $p = q = r = 1/4$ as before.

Lastly, the even quicker result from the half-shift rule: the total translation for *Pnma* is $(b/2 + c/2 + a/2)$, and the half shift is therefore $(b/4 + b/4 + a/4)$. Hence, the origin $(\bar{1})$ is set off by this amount from the point of intersection of the three symmetry planes (Fig. 5.25). The rest of the space group description follows along the lines already discussed.

If we were to be presented with a centred group, for example, *Cmcm*, we could treat it as *Pmcm + C*; *Imma* would be treated as *Pmma + I*. In each case, the totality of symmetry elements must be obtained and scrutinized to ensure that the standard setting is chosen.

In the orthorhombic system, we encounter *d* glide plane symmetry for the first time with space groups *Fdd2* and *Fddd*. They can be drawn up by means of the techniques already described. The reader may care to work out a diagram for *Fddd*, the space group for orthorhombic sulphur S_8, and compare the result with the frontispiece of this book.

Standard settings
The reference work on crystal symmetry is the *International Tables for Crystallography*, Volume A (1983), previously issued as the *International Tables for X-Ray Crystallography*, Volume I (1952, 1965). With certain space groups in the *mmm* class, the first choice of origin is not on a centre of symmetry but on the point group 222, although the alternative origin on $\bar{1}$ is listed. The reason for this choice, in *Pban* for example,

Table 5.4—Orthorhombic space groups

P222	*Pmm2*	*Cmc2₁*	*Pmmm*	*Pbca*
P222₁	*Pmc2₁*	*Ccc2*	*Pnnn*	*Pnma*
P2₁2₁2	*Pcc2*	*Amm2*	*Pccm*	*Cmcm*
P2₁2₁2₁	*Pma2*	*Abm2*	*Pban*	*Cmca*
C222₁	*Pca2₁*	*Ama2*	*Pmma*	*Cmmm*
C222	*Pnc2*	*Aba2*	*Pnna*	*Cccm*
F222	*Pmn2₁*	*Fmm2*	*Pmna*	*Cmma*
I222	*Pba2*	*Fdd2*	*Pcca*	*Ccca*
I2₁2₁2₁	*Pna2₁*	*Imm2*	*Pbam*	*Fmmm*
	Pnn2	*Iba2*	*Pccn*	*Fddd*
	Cmm2	*Ima2*	*Pbcm*	*Immm*
			Pnnm	*Ibam*
			Pmmn	*Ibca*
			Pbcn	*Imma*

is that it leads to a simpler, more symmetrical expression of the coordinates of the general equivalent positions. For our purposes, we shall consider the origin on a centre of symmetry, which leads to a quite correct representation of any space group in this class. The orthorhombic space groups are listed in their standard settings in Table 5.4.

5.4.8 Alternative space group symbols

We shall confine this short discussion to the monoclinic and orthorhombic systems.

Monoclinic space groups

In the monoclinic system, convention fixes the unique axis as y, but the x and z axes may be interchanged leading to different space-group symbols. Consider $C2/c$, related to the right-handed axes x, y, z. From the symbol, we know that the ab plane is C centred, 2 is parallel to b, and c is parallel to the plane ac with a translational component of $c/2$. If we make a transformation to a new cell

$$\mathbf{a}' = -\mathbf{c} \qquad \mathbf{b}' = \mathbf{b} \qquad \mathbf{c}' = \mathbf{a}$$

we must consider a new symbol. The centred plane ab is now $b'c'$, and the new cell is A centred. The glide plane was ac; it is still $a'c'$, but the translational component which was $c/2$ is now $a'/2$. Thus, the space-group symbol referred to the new cell is $A2/a$. Summarizing:

Setting:	**abc**	c̄ba (a′b′c′)
Symbol	$C2/c$	$A2/a$

There is a practical significance in such transformations. While we may correctly recognize the b direction, as it is a symmetry direction, the choice of a and c is not rigid. If we wish to work with a space group in its standard setting, a transformation of the experimentally determined unit cell may be necessary.

Orthorhombic space groups

The orthorhombic system provides more variation, because the directions a, b and c, although determined by the directions of 2 or $\bar{2}$ symmetry, may be chosen in any order, subject to the right-handedness of the sequence of a, b, c.

Consider the standard (**abc**) setting of $Cmcm$. Suppose that the unit cell had been set as **bca**, that is, through the transformation

$$\mathbf{a}' = \mathbf{b}$$

$$\mathbf{b}' = \mathbf{c}$$

$$\mathbf{c}' = \mathbf{a}$$

The centred plane ab is here $a'c'$, that is, B centring, m normal to a is now normal to c'; similarly, the c glide plane is normal to a' with a translation component of $b'/2$, and the third symbol m is normal to b'. Assembling all this information, we arrive at the symbol $Bbmm$. This result can also be obtained by means of a diagram (Fig. 5.26).

5.4.9 Tetragonal space groups

There are many tetragonal space groups, but we shall consider just five, $I\bar{4}$, $P4_2/n$, $P4bn$, $P\dfrac{4}{n}bm$ and $I\dfrac{4_1}{a}md$. We shall call upon some results obtained in Section 5.3.3,

but remembering that we are now in three dimensions.

In space group $I\bar{4}$, we can choose the origin on a point of inversion on the $\bar{4}$ axis, and then treat the space group as $P\bar{4} + I$. With this information, we arrive readily at Fig. 5.27. How are the $\bar{4}$ inversion points at $z = 1/4$ revealed to you?

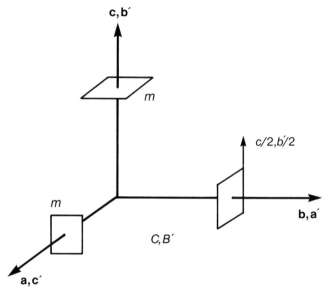

Fig. 5.26—Transformation of the space group setting: the unprimed letters refer to the standard setting **abc** and the symbol *Cmcm*, and the primed symbols relate to the **bca** setting and symbol *Bbmm*.

In the tetragonal system, the projection on to (001) is straightforward; for $I\bar{4}$ it corresponds to plane group $p4$ with $a' = b' = a\sqrt{2}/2$. On (100) the projected plane group is cm, with $a' = b$ and $b' = c$, and on (110), the third important plane in the tetragonal system, it is pm, with $a' = a\sqrt{2}/2$ and $b' = c/2$. Generally, the projected symmetries are of less value in systems of symmetry higher than orthorhombic, and will not be listed further.

The special equivalent positions of $I\bar{4}$ need to be occupied by a molecule of symmetry $\bar{4}$ or 2, or of a symmetry of which $\bar{4}$ or 2 is a sub-group. This situation is realized with pentaerythritol (Fig. 1.1), which crystallizes with this space group and two molecules per unit cell, at $0, 0, 0$ and $\frac{1}{2}, \frac{1}{2}, \frac{1}{2}$.

In the tetragonal space groups related to point groups 4 and $\bar{4}$, we can always select an origin on 4 or $\bar{4}$. Point group $4/m$ is centrosymmetric, and we can choose an origin on a centre of symmetry. As we have already seen, in the presence of translational symmetry, $\bar{1}$ may not lie at the intersection of the symmetry axis and the symmetry plane.

The half-shift rule is not so satisfactory in the higher symmetry systems, but we can proceed by making use of the meaning of the space group symbol, as already discussed. Reference to Fig. 3.13(b) shows that in point group $4/m$, we need successively two operations of 4 followed by m to attain an equivalent position related by $\bar{1}$ to the original point. Consider $P4_2/n$. We may write the following specification:

$$\bar{1} \text{ at } 0, 0, 0$$

$$4_2 \text{ along } [p, q, z]$$

$$n \text{ across } (x, y, r)$$

$$I\bar{4}$$
$$S_4^2$$

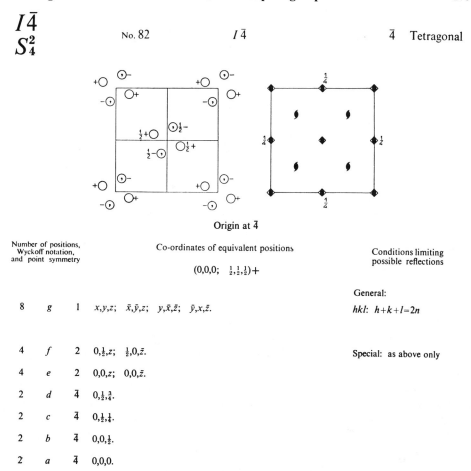

No. 82 $I\bar{4}$ $\bar{4}$ Tetragonal

Origin at $\bar{4}$

Number of positions, Wyckoff notation, and point symmetry			Co-ordinates of equivalent positions $(0,0,0;\ \frac{1}{2},\frac{1}{2},\frac{1}{2})+$	Conditions limiting possible reflections
				General:
8	g	1	$x,y,z;\ \ \bar{x},\bar{y},z;\ \ y,\bar{x},\bar{z};\ \ \bar{y},x,\bar{z}.$	$hkl:\ h+k+l=2n$
4	f	2	$0,\frac{1}{2},z;\ \ \frac{1}{2},0,\bar{z}.$	Special: as above only
4	e	2	$0,0,z;\ \ 0,0,\bar{z}.$	
2	d	$\bar{4}$	$0,\frac{1}{2},\frac{3}{4}.$	
2	c	$\bar{4}$	$0,\frac{1}{2},\frac{1}{4}.$	
2	b	$\bar{4}$	$0,0,\frac{1}{2}.$	
2	a	$\bar{4}$	$0,0,0.$	

Fig. 5.27—Tetragonal space group $I\bar{4}$.

Then, we may obtain

$$x, y, z \xrightarrow{\ 4_2\ } p+q-y,\, q-p+x,\, \tfrac{1}{2}+z \xrightarrow{\ 4_2\ } 2p-x,\, 2q-y,\, z$$

with the $\bar{1}$ and n relations giving

$$\tfrac{1}{2}+2p-x,\, \tfrac{1}{2}+2q-y,\, 2r-z$$

$$\xrightarrow{\ \bar{1}\ } \bar{x},\, \bar{y},\, \bar{z}$$

whence $p = q = 1/4$ and $r = 0$, as in Fig. 5.28(a); the standard diagram is shown in Fig. 5.28(b). It may be noted that the double operation 4_2 is equivalent to a two-fold rotation about the same axis; 2 is a sub-group of 4_2. (The reader should confirm the coordinates from the first 4_2 operation.) The relationships of the x, y, z coordinates in the space groups of types 4, $\bar{4}$ and $4/m$ are shown by the following scheme:

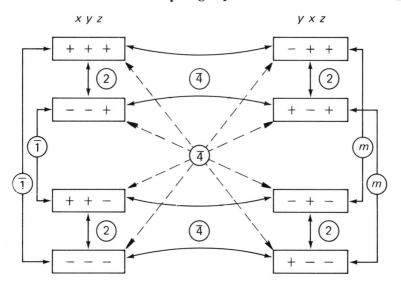

We can see from this scheme that $\bar{4}$ is present in all space groups of type $4/m$, even though in some cases it is coincident with another symmetry element, for example, 4 in $P4/m$. We must remember that $\bar{4}$ is not just a line, it contains a point through which the inversion is carried out; the point may be at the origin, or at some fraction of the cell side c.

In point group $4mm$, the four-fold axis lies at the intersection of two *forms* of m planes, one normal to the x and y axes, and the other normal to [110] and [1$\bar{1}$0], as listed in Table 3.5. We need to consider just one plane in each form, say, m normal to the x axis, and the m normal to [110]. We could draw up a scheme similar to that on page 170, and it can be instructive so to do, and to work out the relationships between the various coordinate triplets. However, let us work with the symbol $P4bm$, together with the appropriate cycle of operations deducible through Fig. 3.18(b). Thus, we write

$$4 \text{ along } [0, 0, z]$$

$$b \text{ across } (p, y, z)$$

$$m \text{ across } (q, q, z)$$

Thus, we have

$$x, y, z \xrightarrow{\quad b \quad} 2p - x, \tfrac{1}{2} + y, z$$
$$\downarrow m$$
$$q - \tfrac{1}{2} - y, q - 2p + x, z$$
$$x, y, z \xrightarrow{\quad 4 \quad} \bar{y}, x, z$$

whence $q = 1/2$ and $p = q/2 = 1/4$, and we can draw up Fig. 5.29. The diagonal reflexion plane normal to [110] always makes equal intersections along the a and b unit cell edges. If we work on $I4cm$ by this method, treating it as $P4cm + I$, the result may at first seem confusing. First, $P4cm$ does not exist; the c glide plane causes the unique axis to be 4_2, and we obtain the orientation of 4_2cm in $I4cm$. The space group

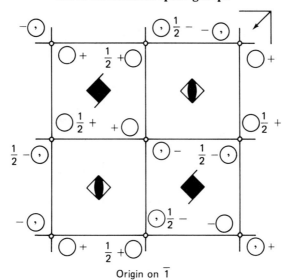

Origin on $\bar{1}$

Fig. 5.28(a)—Space group $P4_2/n$, origin on 1.

$P\,4_2/n$
C^4_{4h}

No. 86 $P\,4_2/n$ $4/m$ Tetragonal

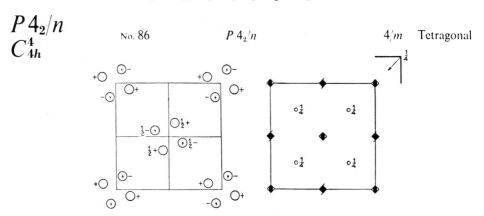

Origin at $\bar{4}$ at $\bar{\frac{1}{4}},\bar{\frac{1}{4}},\bar{\frac{1}{4}}$ from $\bar{1}$ (compare below for alternative origin)

Number of positions, Wyckoff notation, and point symmetry			Co-ordinates of equivalent positions			Conditions limiting possible reflections
						General:
8	g	1	$x,y,z;$ $\quad\bar{x},\bar{y},z;$	$\frac{1}{2}+x,\frac{1}{2}+y,\frac{1}{2}-z;$	$\frac{1}{2}-x,\frac{1}{2}-y,\frac{1}{2}-z;$	$hkl:$ No conditions
			$\bar{y},x,\bar{z};$ $\quad y,\bar{x},\bar{z};$	$\frac{1}{2}-y,\frac{1}{2}+x,\frac{1}{2}+z;$	$\frac{1}{2}+y,\frac{1}{2}-x,\frac{1}{2}+z.$	$hk0:$ $h+k=2n$
						$00l:$ $l=2n$
						Special: as above, plus
4	f	2	$0,0,z;$ $\quad 0,0,\bar{z};$	$\frac{1}{2},\frac{1}{2},\frac{1}{2}+z;$	$\frac{1}{2},\frac{1}{2},\frac{1}{2}-z.$	$hkl:$ $h+k+l=2n$
4	e	2	$0,\frac{1}{2},z;$ $\quad 0,\frac{1}{2},\frac{1}{2}+z;$	$\frac{1}{2},0,\bar{z};$	$\frac{1}{2},0,\frac{1}{2}-z.$	$hkl:$ $l=2n$
4	d	$\bar{1}$	$\frac{1}{4},\frac{1}{4},\frac{3}{4};$ $\quad \frac{3}{4},\frac{3}{4},\frac{3}{4};$	$\frac{1}{4},\frac{3}{4},\frac{1}{4};$	$\frac{3}{4},\frac{1}{4},\frac{1}{4};$	$hkl:$ $h+k=2n$
4	c	$\bar{1}$	$\frac{1}{4},\frac{1}{4},\frac{1}{4};$ $\quad \frac{3}{4},\frac{3}{4},\frac{1}{4};$	$\frac{1}{4},\frac{3}{4},\frac{3}{4};$	$\frac{3}{4},\frac{1}{4},\frac{3}{4}.$	$k+l=2n$ and $(l+h=2n)$
2	b	$\bar{4}$	$0,0,\frac{1}{2};$ $\quad \frac{1}{2},\frac{1}{2},0.$	2 $\quad a$ $\quad \bar{4}$ $\quad 0,0,0;$ $\quad \frac{1}{2},\frac{1}{2},\frac{1}{2}.$		$hkl:$ $h+k+l=2n$ (a and b)

Fig. 5.28(b)—Space group $P4_2/n$, origin (standard) on $\bar{4}$.

$P4bm$
C_{4v}^{2}

No. 100 $P\,4\,b\,m$ $4\,m\,m$ Tetragonal

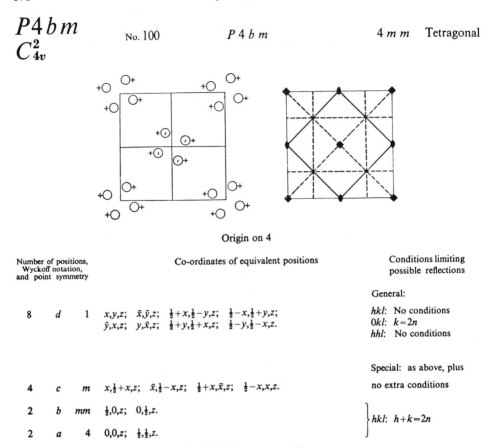

Origin on 4

Number of positions, Wyckoff notation, and point symmetry			Co-ordinates of equivalent positions	Conditions limiting possible reflections

General:

| 8 | d | 1 | $x,y,z;\quad \bar{x},\bar{y},z;\quad \tfrac{1}{2}+x,\tfrac{1}{2}-y,z;\quad \tfrac{1}{2}-x,\tfrac{1}{2}+y,z;$
 $\bar{y},x,z;\quad y,\bar{x},z;\quad \tfrac{1}{2}+y,\tfrac{1}{2}+x,z;\quad \tfrac{1}{2}-y,\tfrac{1}{2}-x,z.$ | hkl: No conditions
 $0kl$: $k=2n$
 hhl: No conditions |

Special: as above, plus

no extra conditions

| 4 | c | m | $x,\tfrac{1}{2}+x,z;\quad \bar{x},\tfrac{1}{2}-x,z;\quad \tfrac{1}{2}+x,\bar{x},z;\quad \tfrac{1}{2}-x,x,z.$ | |

| 2 | b | mm | $\tfrac{1}{2},0,z;\quad 0,\tfrac{1}{2},z.$ | hkl: $h+k=2n$ |

| 2 | a | 4 | $0,0,z;\quad \tfrac{1}{2},\tfrac{1}{2},z.$ | |

Fig. 5.29—Space group $P4bm$.

diagram can be completed fully from this orientation, and then the origin changed to the standard setting on 4. The second point is that if $I4cm$ was named according to the rules of precedence, it would be $I4bm$, and the standard setting is then given immediately by $P4bm + I$. Departures from the rules of precedence are made for certain reasons: $P2_1/c$ is used instead of $P2_1/a$ because $P2_1/c + C \to C2/c$, whereas $P2_1/a + C \to C2/m$. Among the orthorhombic space groups, $Ibca$ is used instead of $Ibaa$; $Pbca + I \to Ibca$. The logic is always there, although sometimes it may be a little difficult to discover.

The reader may like to confirm geometrically, or otherwise, that reflexion of a point x, y, z across the m plane (q, q, z) normal to [110] gives rise to the symmetry-related point $q - y, q - x, z$; this result was used above.

It is possible to work out similar schemes for all classes of tetragonal space groups, and indeed for all space groups. The half-shift rule is not satisfactory in the high symmetry systems. The reader should, perhaps, work through some other tetragonal space groups, say, $P4_12_12$ and $P\bar{4}2c$, so as to realize the full power of the method given above. In $P4_12_12$, the 4_1 axis causes the 2_1 axes parallel to x and y to be separated by $c/4$, and similarly for the diagonal two-fold axes. Since the latter axes interleave the first set in z, they will be separated from them by $1/8$ and $3/8$. The origin is not fixed uniquely in z by the four-fold axis, so we can choose either the 2_1

or the 2 (standard) to be at $z = 0$. In $P\bar{4}2c$, the element $\bar{4}$ is along the z axis, that is $[0, 0, z]$, but is further located by the point of inversion on the axis. Thus, we may take $\bar{4}$ at $0, 0, 0$ so that x, y, z is transformed through this operation to $y, \bar{x}, \bar{z}$.

Our final class of tetragonal space groups is $\dfrac{4}{m} mm$. In this class, the unique axis is set along $[0, 0, z]$, which may result in the centre of symmetry being displaced from the origin. It is always possible to reset the group with $\bar{1}$ at $0, 0, 0$, if necessary.

We will consider $P\dfrac{4}{n} bm$. The four-fold axis is along $[0, 0, z]$ with the perpendicular symmetry plane (n) across (x, y, r). Then, $\bar{1}$ can be set at $p, q, 0$; it is permissible to fix the third coordinate of either $\bar{1}$ or n as zero, because the origin is not defined with respect to z by the other symmetry elements. Where this symmetry plane is m, $p = q = 0$, whereas if it is a glide plane, the values of p and q will depend on the nature of the glide plane.

We write, for $P\dfrac{4}{n} bm$

$$4 \text{ along } [0, 0, z]$$

$$n \text{ across } (x, y, r)$$

$$\bar{1} \text{ at } p, q, 0$$

$$b \text{ across } (s, y, z)$$

$$m \text{ across } (t, t, z)$$

where p, q, r, s and t are to be determined; we have two cycles to assist us here:

$$x, y, z \xrightarrow{\;\;4^2\;\;} \bar{x}, \bar{y}, z$$
$$\downarrow n$$
$$\tfrac{1}{2} - x, \tfrac{1}{2} - y, 2r - z$$
$$\xrightarrow{\;\;\bar{1}\;\;} 2p - x, 2q - y, \bar{z}$$

whence, $p = q = 1/4$ and $r = 0$. Next, and using these results,

$$x, y, z \xrightarrow{\;\;4\;\;} \bar{y}, x, z$$
$$\downarrow n$$
$$\tfrac{1}{2} - y, \tfrac{1}{2} + x, \bar{z}$$
$$\downarrow b$$
$$2s - \tfrac{1}{2} + y, x, \bar{z}$$
$$\downarrow m$$
$$t - x, t - 2s + \tfrac{1}{2} - y, \bar{z}$$
$$\xrightarrow{\;\;\bar{1}\;\;} \tfrac{1}{2} - x, \tfrac{1}{2} - y, \bar{z}$$

whence, $t = 1/2$, and $s = t/2 = 1/4$. We now have the full orientation for $P\dfrac{4}{n} bm$, and when put together in the usual way it leads to Fig. 5.30.

Tetragonal　　4/*m m m*　　　　　　$P\,4/n\,2/b\,2/m$　　　　No. 125　　$P4/n\,b\,m$

$$D_{4h}^{3}$$

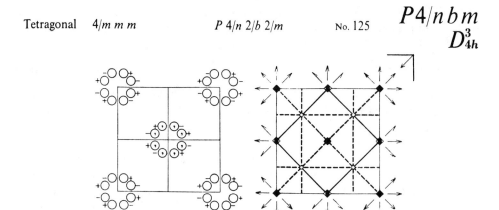

Origin at 422, at $\bar{1},\bar{1},0$ from centre ($2/m$) (compare next page for alternative origin)

Number of positions, Wyckoff notation, and point symmetry			Co-ordinates of equivalent positions			Conditions limiting possible reflections
						General:
16	*n*	1	$x,y,z;$　$\bar{x},\bar{y},z;$　$\tfrac{1}{2}+x,\tfrac{1}{2}+y,\bar{z};$　$\tfrac{1}{2}-x,\tfrac{1}{2}-y,\bar{z};$			hkl: No conditions
			$\bar{x},y,\bar{z};$　$x,\bar{y},\bar{z};$　$\tfrac{1}{2}+x,\tfrac{1}{2}-y,z;$　$\tfrac{1}{2}-x,\tfrac{1}{2}+y,z;$			$hk0$: $h+k=2n$
			$\bar{y},x,z;$　$y,\bar{x},z;$　$\tfrac{1}{2}-y,\tfrac{1}{2}+x,\bar{z};$　$\tfrac{1}{2}+y,\tfrac{1}{2}-x,\bar{z};$			$0kl$: $k=2n$
			$y,x,\bar{z};$　$\bar{y},\bar{x},\bar{z};$　$\tfrac{1}{2}+y,\tfrac{1}{2}+x,z;$　$\tfrac{1}{2}-y,\tfrac{1}{2}-x,z.$			hhl: No conditions
						Special: as above, plus
8	*m*	*m*	$x,\tfrac{1}{2}+x,z;$　$\bar{x},\tfrac{1}{2}-x,z;$　$\tfrac{1}{2}+x,x,\bar{z};$　$\tfrac{1}{2}-x,\bar{x},\bar{z};$			no extra conditions
			$x,\tfrac{1}{2}-x,\bar{z};$　$\bar{x},\tfrac{1}{2}+x,\bar{z};$　$\tfrac{1}{2}+x,\bar{x},z;$　$\tfrac{1}{2}-x,x,z.$			
8	*l*	2	$x,0,\tfrac{1}{2};$　$\bar{x},0,\tfrac{1}{2};$　$\tfrac{1}{2}+x,\tfrac{1}{2},\tfrac{1}{2};$　$\tfrac{1}{2}-x,\tfrac{1}{2},\tfrac{1}{2};$			
			$0,x,\tfrac{1}{2};$　$0,\bar{x},\tfrac{1}{2};$　$\tfrac{1}{2},\tfrac{1}{2}+x,\tfrac{1}{2};$　$\tfrac{1}{2},\tfrac{1}{2}-x,\tfrac{1}{2}.$			
8	*k*	2	$x,0,0;$　$\bar{x},0,0;$　$\tfrac{1}{2}+x,\tfrac{1}{2},0;$　$\tfrac{1}{2}-x,\tfrac{1}{2},0;$			
			$0,x,0;$　$0,\bar{x},0;$　$\tfrac{1}{2},\tfrac{1}{2}+x,0;$　$\tfrac{1}{2},\tfrac{1}{2}-x,0.$			
8	*j*	2	$x,x,\tfrac{1}{2};$　$\bar{x},\bar{x},\tfrac{1}{2};$　$\tfrac{1}{2}+x,\tfrac{1}{2}+x,\tfrac{1}{2};$　$\tfrac{1}{2}-x,\tfrac{1}{2}-x,\tfrac{1}{2};$			hkl: $h+k=2n$
			$x,\bar{x},\tfrac{1}{2};$　$\bar{x},x,\tfrac{1}{2};$　$\tfrac{1}{2}+x,\tfrac{1}{2}-x,\tfrac{1}{2};$　$\tfrac{1}{2}-x,\tfrac{1}{2}+x,\tfrac{1}{2}.$			
8	*i*	2	$x,x,0;$　$\bar{x},\bar{x},0;$　$\tfrac{1}{2}+x,\tfrac{1}{2}+x,0;$　$\tfrac{1}{2}-x,\tfrac{1}{2}-x,0;$			
			$x,\bar{x},0;$　$\bar{x},x,0;$　$\tfrac{1}{2}+x,\tfrac{1}{2}-x,0;$　$\tfrac{1}{2}-x,\tfrac{1}{2}+x,0.$			
4	*h*	*mm*	$0,\tfrac{1}{2},z;$　$0,\tfrac{1}{2},\bar{z};$　$\tfrac{1}{2},0,z;$　$\tfrac{1}{2},0,\bar{z}.$			
4	*g*	4	$0,0,z;$　$0,0,\bar{z};$　$\tfrac{1}{2},\tfrac{1}{2},z;$　$\tfrac{1}{2},\tfrac{1}{2},\bar{z}.$			
4	*f*	$2/m$	$\tfrac{1}{4},\tfrac{1}{4},\tfrac{1}{2};$　$\tfrac{3}{4},\tfrac{3}{4},\tfrac{1}{2};$　$\tfrac{1}{4},\tfrac{3}{4},\tfrac{1}{2};$　$\tfrac{3}{4},\tfrac{1}{4},\tfrac{1}{2}.$			hkl: $h,k=2n$
4	*e*	$2/m$	$\tfrac{1}{4},\tfrac{1}{4},0;$　$\tfrac{3}{4},\tfrac{3}{4},0;$　$\tfrac{1}{4},\tfrac{3}{4},0;$　$\tfrac{3}{4},\tfrac{1}{4},0.$			

Fig. 5.30—Space group $P\dfrac{4}{n}bm$.

Several space groups in this class have $\bar{4}$ symmetry parallel to the z axis, though not explicit in the symbol. In these space groups, $\bar{4}$ is chosen as the origin 0, 0, 0 in the standard settings.

Two groups in this class, $I\dfrac{4_1}{a}md$ and $I\dfrac{4_1}{a}cd$, related to $I4_1md$ and $I4_1cd$ respectively, have diagonal d glides, with translations of $(a\pm b\pm c)/4$. They can be treated by the above methods, but the standard settings place the implicit $\bar{4}$ at $0, 0, 0$. Unlike the values $+\frac{1}{2}$ and $-\frac{1}{2}$ along a given axis, which are Bravais equivalent positions (that is, they differ by a Bravais unit cell translation), the inversion points along 4 occur in these groups at $\pm\frac{1}{4}$, and are not of the same character. Thus, it is necessary to ensure that the directions of the $\frac{1}{4}$ translations are applied to x, y and z in the same sense, in the d symmetry operation, so as to bring the point finally to the position from where $\bar{1}$ will return it to x, y, z. Let us consider $I \dfrac{4_1}{a} md$:

$$4_1 \text{ along } [0, 0, z]$$

$$a \text{ across } (x, y, r)$$

$$\bar{1} \text{ at } p, q, 0$$

$$m \text{ across } (s, y, z)$$

$$d \text{ across } (t, t, z)$$

Then, we have

$$x, y, z \xrightarrow{\quad 4_1^2 \quad} \bar{x}, \bar{y}, \tfrac{1}{2} + z$$

$$\downarrow a$$

$$\tfrac{1}{2} - x, \bar{y}, 2r - \tfrac{1}{2} - z$$

$$\xrightarrow{\quad \bar{1} \quad} 2p - x, 2q - y, \bar{z}$$

whence $p = 1/4$, $q = 0$ and $r = 1/4$. Again, and using these results,

$$x, y, z \xrightarrow{\quad 4_1 \quad} \bar{y}, x, \tfrac{1}{4} + z$$

$$\downarrow a$$

$$\tfrac{1}{2} - y, x, \tfrac{1}{4} - z$$

$$\downarrow m$$

$$2s - \tfrac{1}{2} + y, x, \tfrac{1}{4} - z$$

$$\downarrow d(+a/4 - b/4 - c/4)$$

$$\tfrac{1}{4} + t - x, -\tfrac{1}{4} + t - 2s + \tfrac{1}{2} - y, \bar{z}$$

$$\xrightarrow{\quad \bar{1} \quad} \tfrac{1}{2} - x, \bar{y}, \bar{z}$$

whence $s = t = 1/4$. Fig. 5.31 shows the standard setting of this space group. We can see that our analysis has placed a centre of symmetry at $\frac{1}{4}, \frac{1}{4}, 0$, and the diagram can easily be re-oriented to give the standard setting by moving the origin to $\bar{4}$ (at $\frac{1}{4}, \frac{1}{4}, \frac{1}{8}$), which adjusts the vertical heights of the a glide plane and the centre of symmetry by $+\frac{1}{8}$ each.

Table 5.5 lists the space groups in the tetragonal system in their standard settings.

5.4.10 Trigonal, hexagonal and cubic space groups

The space groups in these three systems can be worked through by methods similar to those discussed in the previous sections of this chapter, but they can be a little more tricky.

Tetragonal $4/m\,m\,m$ $I\,4_1/a\,2/m\,2/d$ No. 141 $I4_1/a\,m\,d$
D_{4h}^{19}

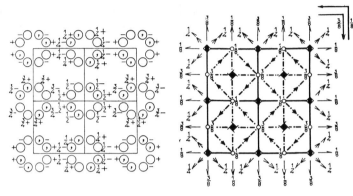

Origin at $\bar{4}m2$, at $0,\tfrac{1}{4},\bar{\tfrac{1}{8}}$ from centre $(2/m)$ (compare next page for alternative origin)

Number of positions, Wyckoff notation, and point symmetry			Co-ordinates of equivalent positions $(0,0,0;\ \tfrac{1}{2},\tfrac{1}{2},\tfrac{1}{2})+$				Conditions limiting possible reflections
							General:
32	i	1	$x,y,z;$	$\bar{x},\bar{y},z;$	$x,\tfrac{1}{2}+y,\tfrac{1}{4}-z;$	$\bar{x},\tfrac{1}{2}-y,\tfrac{1}{4}-z;$	$hkl:\ h+k+l=2n$
			$\bar{x},y,z;$	$x,\bar{y},z;$	$\bar{x},\tfrac{1}{2}+y,\tfrac{1}{4}-z;$	$x,\tfrac{1}{2}-y,\tfrac{1}{4}-z;$	$hk0:\ h,(k)=2n$
			$y,x,\bar{z};$	$\bar{y},\bar{x},\bar{z};$	$y,\tfrac{1}{2}+x,\tfrac{1}{4}+z;$	$\bar{y},\tfrac{1}{2}-x,\tfrac{1}{4}+z;$	$0kl:\ (k+l=2n)$
			$\bar{y},x,\bar{z};$	$y,\bar{x},\bar{z};$	$\bar{y},\tfrac{1}{2}+x,\tfrac{1}{4}+z;$	$y,\tfrac{1}{2}-x,\tfrac{1}{4}+z.$	$hhl:\ (l=2n);\ 2h+l=4n$
							Special: as above, plus
16	h	m	$0,x,z;$	$0,\bar{x},z;$	$0,\tfrac{1}{2}+x,\tfrac{1}{4}-z;$	$0,\tfrac{1}{2}-x,\tfrac{1}{4}-z;$	no extra conditions
			$x,0,\bar{z};$	$\bar{x},0,\bar{z};$	$x,\tfrac{1}{2},\tfrac{1}{4}+z;$	$\bar{x},\tfrac{1}{2},\tfrac{1}{4}+z.$	
16	g	2	$x,x,0;$	$\bar{x},\bar{x},0;$	$x,\tfrac{1}{2}+x,\tfrac{1}{4};$	$\bar{x},\tfrac{1}{2}-x,\tfrac{1}{4};$	$hkl:\ 2k+l=2n+1\ \text{or}\ 4n$
			$x,\bar{x},0;$	$\bar{x},x,0;$	$x,\tfrac{1}{2}-x,\tfrac{1}{4};$	$\bar{x},\tfrac{1}{2}+x,\tfrac{1}{4}.$	
16	f	2	$x,\tfrac{1}{4},\tfrac{1}{8};$	$\bar{x},\tfrac{1}{4},\tfrac{1}{8};$	$x,\tfrac{3}{4},\tfrac{1}{8};$	$\bar{x},\tfrac{3}{4},\tfrac{1}{8};$	$hkl:\ l,(h+k)=2n+1,\ \text{or}$
			$\tfrac{1}{4},x,\tfrac{7}{8};$	$\tfrac{1}{4},\bar{x},\tfrac{7}{8};$	$\tfrac{3}{4},x,\tfrac{7}{8};$	$\tfrac{3}{4},\bar{x},\tfrac{7}{8}.$	$h,k,(l)=2n$
8	e	mm	$0,0,z;$	$0,0,\bar{z};$	$0,\tfrac{1}{2},\tfrac{1}{4}+z;$	$0,\tfrac{1}{2},\tfrac{1}{4}-z.$	$hkl:\ 2k+l=2n+1\ \text{or}\ 4n$
8	d	$2/m$	$0,\tfrac{1}{4},\tfrac{5}{8};$	$0,\tfrac{3}{4},\tfrac{5}{8};$	$\tfrac{1}{4},0,\tfrac{3}{8};$	$\tfrac{3}{4},0,\tfrac{3}{8}.$	$hkl:\ l,(h+k)=2n+1,\ \text{or}$
8	c	$2/m$	$0,\tfrac{1}{4},\tfrac{1}{8};$	$0,\tfrac{3}{4},\tfrac{1}{8};$	$\tfrac{1}{4},0,\tfrac{7}{8};$	$\tfrac{3}{4},0,\tfrac{7}{8}.$	$[h,k,(l)=2n;$ $h+k+l=4n]$
4	b	$\bar{4}m2$	$0,0,\tfrac{1}{2};$	$0,\tfrac{1}{2},\tfrac{3}{4}.$			$hkl:\ 2k+l=2n+1\ \text{or}\ 4n$
4	a	$\bar{4}m2$	$0,0,0;$	$0,\tfrac{1}{2},\tfrac{1}{4}.$			

Fig. 5.31—Space group $I\dfrac{4_1}{a}md$; the origin used to derive the diagram is the 4_1 axis shown here at $\tfrac{1}{4},\tfrac{3}{4},z$, for example.

In the trigonal system, we have to consider the three-fold symmetry with both the P (hexagonal) lattice and the R (rhombohedral, observe setting) lattice. The rhombohedral lattice may be referred to hexagonal axes in a hexagonal triply primitive unit cell; Fig. 5.32 shows diagrams of space group $R\bar{3}c$, on both rhombohedral and hexagonal axes.

Table 5.5—Tetragonal space groups

$P4$	$P\bar{4}$	$P4/m$	$P422$	$P4mm$	$P\bar{4}2m$	$P\dfrac{4}{m}mm$
$P4_1$	$I\bar{4}$	$P4_2/m$	$P42_12$	$P4bm$	$P\bar{4}2c$	$P\dfrac{4}{m}cc$
$P4_2$		$P4/n$	$P4_122$	$P4_2cm$	$P\bar{4}2_1m$	$P\dfrac{4}{n}bm$
$P4_3$		$P4_2/n$	$P4_12_12$	$P4_2nm$	$P\bar{4}2_1c$	$P\dfrac{4}{n}nc$
$I4$		$I4/m$	$P4_222$	$P4cc$	$P\bar{4}m2$	$P\dfrac{4}{m}bm$
$I4_1$		$I4_1/a$	$P4_22_12$	$P4nc$	$P\bar{4}c2$	$P\dfrac{4}{m}nc$
			$P4_322$	$P4_2mc$	$P\bar{4}b2$	$P\dfrac{4}{n}mm$
			$P4_32_12$	$P4_2bc$	$P\bar{4}n2$	$P\dfrac{4}{n}cc$
			$I422$	$I4mm$	$I\bar{4}m2$	$P\dfrac{4_2}{m}mc$
			$I4_122$	$I4cm$	$I\bar{4}c2$	$P\dfrac{4_2}{m}cm$
				$I4_1md$	$I\bar{4}2m$	$P\dfrac{4_2}{n}bc$
				$I4_2cd$	$I\bar{4}2d$	$P\dfrac{4_2}{n}nm$
						$P\dfrac{4_2}{m}bc$
						$P\dfrac{4_2}{m}nm$
						$P\dfrac{4_2}{n}mc$
						$P\dfrac{4_2}{n}cm$
						$I\dfrac{4}{m}mm$
						$I\dfrac{4}{m}cm$
						$I\dfrac{4_1}{a}md$
						$I\dfrac{4_1}{a}cd$

$R\bar{3}c$
D_{3d}^6

No. 167 $R\,\bar{3}\,2/c$ $\bar{3}\,m$ Trigonal

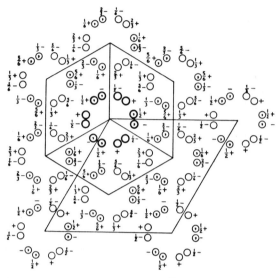

Origin at centre ($\bar{3}$)

Number of positions, Wyckoff notation, and point symmetry			Co-ordinates of equivalent positions	Conditions limiting possible reflections
			(1) RHOMBOHEDRAL AXES:	
				General:
12	f	1	$x,y,z;\quad z,x,y;\quad y,z,x;\quad \bar{x},\bar{y},\bar{z};\quad \bar{z},\bar{x},\bar{y};\quad \bar{y},\bar{z},\bar{x};$ $\tfrac{1}{2}+y,\tfrac{1}{2}+x,\tfrac{1}{2}+z;\quad \tfrac{1}{2}+z,\tfrac{1}{2}+y,\tfrac{1}{2}+x;\quad \tfrac{1}{2}+x,\tfrac{1}{2}+z,\tfrac{1}{2}+y;$ $\tfrac{1}{2}-y,\tfrac{1}{2}-x,\tfrac{1}{2}-z;\quad \tfrac{1}{2}-z,\tfrac{1}{2}-y,\tfrac{1}{2}-x;\quad \tfrac{1}{2}-x,\tfrac{1}{2}-z,\tfrac{1}{2}-y.$	hkl: No conditions hhl: l=2n
				Special: as above, plus
6	e	2	$x,\tfrac{1}{2}-x,\tfrac{1}{4};\quad \tfrac{1}{2}-x,\tfrac{1}{4},x;\quad \tfrac{1}{4},x,\tfrac{1}{2}-x;$ $\bar{x},\tfrac{1}{2}+x,\tfrac{3}{4};\quad \tfrac{1}{2}+x,\tfrac{3}{4},\bar{x};\quad \tfrac{3}{4},\bar{x},\tfrac{1}{2}+x.$	no extra conditions
6	d	$\bar{1}$	$\tfrac{1}{2},0,0;\quad 0,\tfrac{1}{2},0;\quad 0,0,\tfrac{1}{2};\quad 0,\tfrac{1}{2},\tfrac{1}{2};\quad \tfrac{1}{2},0,\tfrac{1}{2};\quad \tfrac{1}{2},\tfrac{1}{2},0.$	
4	c	3	$x,x,x;\quad \bar{x},\bar{x},\bar{x};\quad \tfrac{1}{2}+x,\tfrac{1}{2}+x,\tfrac{1}{2}+x;\quad \tfrac{1}{2}-x,\tfrac{1}{2}-x,\tfrac{1}{2}-x.$	
2	b	$\bar{3}$	$0,0,0;\quad \tfrac{1}{2},\tfrac{1}{2},\tfrac{1}{2}.$	hkl: h+k+l=2n
2	a	32	$\tfrac{1}{4},\tfrac{1}{4},\tfrac{1}{4};\quad \tfrac{3}{4},\tfrac{3}{4},\tfrac{3}{4}.$	

Fig. 5.32—Trigonal space group $R\bar{3}c$; the pattern is referred to both the rhombohedral R cell and the triply primitive hexagonal cell (see also Fig. 4.27(a) and (b)).

The matrix for a three-fold anticlockwise rotation about the z axis is given in Appendix 5 (A5.12), from which we see that a point $x,\ y,\ z$ is transformed to $\bar{y},\ x-y,\ z$, if referred to hexagonal axes, with 3 along the z axis. On rhombohedral axes, the three-fold axis is along [111], and $x,\ y,\ z$ transforms cyclically to $z,\ x,\ y$ by a right-handed rotation of 120°.

In the hexagonal system, the point related by a six-fold rotation about z can be obtained from (A5.12), together with a coincident two-fold rotation, such that

$$2 \cdot 3^2 \equiv 6 \tag{5.7}$$

Trigonal $\bar{3}\,m$ $R\,\bar{3}\,2/c$ No. 167 $R\bar{3}c$
D_{3d}^6
(*continued*)

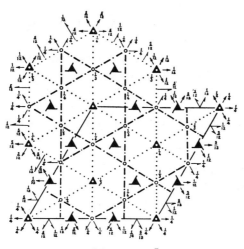

Origin at centre ($\bar{3}$)

Number of positions, Wyckoff notation, and point symmetry	Co-ordinates of equivalent positions	Conditions limiting possible reflections

(2) HEXAGONAL AXES:

$$(0,0,0;\ \tfrac{2}{3},\tfrac{1}{3},\tfrac{1}{3};\ \tfrac{1}{3},\tfrac{2}{3},\tfrac{2}{3})+$$

General:

36 f 1 $x,y,z;$ $\quad \bar{y},x-y,z;$ $\quad y-x,\bar{x},z;$
$\bar{x},\bar{y},\bar{z};$ $\quad y,y-x,\bar{z};$ $\quad x-y,x,\bar{z};$
$\bar{y},\bar{x},\tfrac{1}{2}+z;$ $\quad x,x-y,\tfrac{1}{2}+z;$ $\quad y-x,y,\tfrac{1}{2}+z;$
$y,x,\tfrac{1}{2}-z;$ $\quad \bar{x},y-x,\tfrac{1}{2}-z;$ $\quad x-y,\bar{y},\tfrac{1}{2}-z.$

$hkil:\quad -h+k+l=3n$
$hh2\bar{h}l:\quad (l=3n)$
$hh0l:\quad (h+l=3n);\quad l=2n$

Special: as above, plus

18 e 2 $x,0,\tfrac{1}{4};\quad 0,x,\tfrac{1}{4};\quad \bar{x},\bar{x},\tfrac{1}{4};\quad \bar{x},0,\tfrac{3}{4};\quad 0,\bar{x},\tfrac{3}{4};\quad x,x,\tfrac{3}{4}.$

no extra conditions

18 d $\bar{1}$ $\tfrac{1}{2},0,0;\quad 0,\tfrac{1}{2},0;\quad \tfrac{1}{2},\tfrac{1}{2},0;\quad \tfrac{1}{2},0,\tfrac{1}{2};\quad 0,\tfrac{1}{2},\tfrac{1}{2};\quad \tfrac{1}{2},\tfrac{1}{2},\tfrac{1}{2}.$

12 c 3 $0,0,z;\quad 0,0,\bar{z};\quad 0,0,\tfrac{1}{2}+z;\quad 0,0,\tfrac{1}{2}-z.$

6 b $\bar{3}$ $0,0,0;\quad 0,0,\tfrac{1}{2}.$

6 a 32 $0,0,\tfrac{1}{4};\quad 0,0,\tfrac{3}{4}.$

$hkil:\quad l=2n$

and x,y,z is transformed as follows:

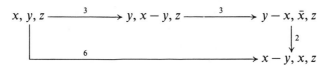

The general orientation of symmetry elements follows Table 3.5, the relative orientation for any space group depending on the nature of the symmetry elements present.

In class 32, we find that $P321$ and $P312$, for example, are different. Similarly, $P3m1$ and $P31m$ also give rise to differing spatial arrangements, but the corresponding $R32$ and $R3m$ are unique, with 2 or $\bar{2}$ along the x, y and u axes.

In the trigonal and hexagonal space groups, the only type of glide plane to arise is the c glide, but screw axes can be 3_1 and 3_2 in the trigonal system, and 6_1, 6_2, 6_3, 6_4 and 6_5 in the hexagonal system.

In the cubic system, Table 3.5 shows that the first position in the symbol refers to the x, y and z directions, the second position to $\langle 111 \rangle$ and the third to $\langle 110 \rangle$. We now need the transformations for rotation about [111] and [110] in the cubic system.

In space group $P23$, the simplest cubic space group, the [111] rotation has, for three points, the same cyclic transformation as in $R3$, namely

$$x, y, z \xrightarrow{\quad 3_{[111]} \quad} z, x, y \xrightarrow{\quad 3_{[111]} \quad} y, z, x$$

The remaining nine points can be generated by the effect of the two-fold rotations about x, y and z. These rotations are implicit in a group of four three-fold axes at $\cos^{-1} 1/\sqrt{3}$ to x, y and z. Thus, we obtain from z, x, y, for example,

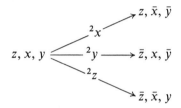

It is because of these relations in the cubic system that point group type R refers to 23 (Table 3.8). The matrices for 3-fold rotation about [111] and [$\bar{1}1\bar{1}$] are

$$\begin{vmatrix} 0 & 0 & 1 \\ 1 & 0 & 0 \\ 0 & 1 & 0 \end{vmatrix} \quad \text{and} \quad \begin{vmatrix} 0 & 0 & 1 \\ \bar{1} & 0 & 0 \\ 0 & \bar{1} & 0 \end{vmatrix}$$

respectively, so that it is easy to show that $3^2_{[\bar{1}1\bar{1}]} \cdot 3_{[111]} \equiv 2_z$.

The rotation about [110], as exists in $P432$, for example, can be obtained by a combination. From the stereogram for point group $m3m$ (Fig. 3.21(e)), we see that a right-handed rotation of 90° about the x axis followed by a right-handed three-fold rotation about [111] is equivalent to a two-fold rotation about [110]. Thus, we have, in matrix notation

$$\underset{3_{[111]}}{\begin{vmatrix} 0 & 0 & 1 \\ 1 & 0 & 0 \\ 0 & 1 & 0 \end{vmatrix}} \quad \underset{4_x}{\begin{vmatrix} 1 & 0 & 0 \\ 0 & 0 & \bar{1} \\ 0 & 1 & 0 \end{vmatrix}} = \underset{2_{[110]}}{\begin{vmatrix} 0 & 1 & 0 \\ 1 & 0 & 0 \\ 0 & 0 & \bar{1} \end{vmatrix}}$$

or

$$x, y, z \xrightarrow{\quad 2_{[110]} \quad} y, x, \bar{z}$$

Diagrams for the cubic space groups are complicated, because of the inclined three-fold axes, and we shall not consider them here. The reader is referred to the *International*

Table 5.6—Trigonal, hexagonal and cubic space groups

Trigonal						
$P3$	$P\bar{3}$	$P312$	$P3m1$	$P\bar{3}1m$		
$P3_1$	$R\bar{3}$	$P321$	$P31m$	$P\bar{3}1c$		
$P3_2$		$P3_112$	$P3c1$	$P\bar{3}m1$		
$R3$		$P3_121$	$P31c$	$P\bar{3}c1$		
		$P3_212$	$R3m$	$R\bar{3}m$		
		$P3_221$	$R3c$	$R\bar{3}c$		
		$R32$				

Hexagonal						
$P6$	$P\bar{6}$	$P6/m$	$P622$	$P6mm$	$P\bar{6}m2$	$P\dfrac{6}{m}mm$
$P6_1$		$P6_3/m$	$P6_122$	$P6cc$	$P\bar{6}c2$	$P\dfrac{6}{m}cc$
$P6_5$			$P6_522$	$P6_3cm$	$P\bar{6}2m$	$P\dfrac{6}{m}cm$
$P6_2$			$P6_222$	$P6_3mc$	$P\bar{6}2c$	$P\dfrac{6}{m}mc$
$P6_4$			$P6_422$			
$P6_3$			$P6_322$			

Cubic						
$P23$	$Pm3$	$P432$	$P\bar{4}3m$	$Pm3m$		
$F23$	$Pn3$	$P4_232$	$F\bar{4}3m$	$Pn3n$		
$I23$	$Fm3$	$F432$	$I\bar{4}3m$	$Pm3n$		
$P2_13$	$Fd3$	$F4_132$	$P\bar{4}3n$	$Pn3m$		
$I2_13$	$Im3$	$I432$	$F\bar{4}3c$	$Fm3m$		
	$Pa3$	$P4_332$	$I\bar{4}3d$	$Fm3c$		
	$Ia3$	$P4_132$		$Fd3m$		
		$I4_132$		$Fd3c$		
				$Im3m$		
				$Ia3d$		

Tables for Crystallography, Volume A (1983). The trigonal, hexagonal and cubic space groups are listed in Table 5.6.

5.4.11 Matrix representation of space-group symmetry operations

In Section 3.9, we considered a matrix representation of symmetry operations. In (3.4), **t** is zero in point groups, but in space groups it takes on values that depend upon both the nature of the symmetry elements, and their relative orientations with respect to the origin.

Consider space group $P2_1/c$. As before, we take the origin on $\bar{1}$, with 2_1 along $[p, y, r]$ and c across (x, q, x). Then, we have

$$
\begin{bmatrix} 1 & 0 & 0 \\ 0 & \bar{1} & 0 \\ 0 & 0 & 1 \end{bmatrix} + \begin{bmatrix} 0 \\ 2q \\ \frac{1}{2} \end{bmatrix} \cdot \begin{bmatrix} \bar{1} & 0 & 0 \\ 0 & 1 & 0 \\ 0 & 0 & \bar{1} \end{bmatrix} + \begin{bmatrix} 2p \\ \frac{1}{2} \\ 2r \end{bmatrix} = \begin{bmatrix} \bar{1} & 0 & 0 \\ 0 & \bar{1} & 0 \\ 0 & 0 & \bar{1} \end{bmatrix} + \begin{bmatrix} 0 \\ 0 \\ 0 \end{bmatrix} \qquad (5.8)
$$

$$\quad \mathbf{R_2} \qquad\quad \mathbf{t_2} \qquad\qquad \mathbf{R_1} \qquad\qquad \mathbf{t_1} \qquad\quad \mathbf{R_3} \qquad\quad \mathbf{t_3}$$

The matrices $\mathbf{R}_1$, $\mathbf{R}_2$ and $\mathbf{R}_3$ are identical with those used for point group $2/m$ (3.8), and are manupulated in the same way, that is,

$$\mathbf{R}_2 \cdot \mathbf{R}_1 \equiv \mathbf{R}_3 \tag{5.9}$$

The translation vectors are added:

$$\mathbf{t}_2 + \mathbf{t}_1 \equiv \mathbf{t}_3 \tag{5.10}$$

Thus, it follows that $p = 0$ and $q = r = 1/4$, as determined in Section 5.3.4 and because (5.6) holds. We can accept that $q = +\frac{1}{4}$ is equivalent to $q = -\frac{1}{4}$: alternatively, we could take a negative component for the translations in both the 2_1 and c symmetry operations.

We look at $P4bm$ as a second example, with 4 along $[0, 0, z]$, b across (p, y, z) and m across (q, q, z). We can write

$$
\begin{vmatrix} 0 & \bar{1} & 0 \\ \bar{1} & 0 & 0 \\ 0 & 0 & 1 \end{vmatrix} + \begin{vmatrix} q \\ q \\ 0 \end{vmatrix} \cdot \begin{vmatrix} \bar{1} & 0 & 0 \\ 0 & 1 & 0 \\ 0 & 0 & 1 \end{vmatrix} + \begin{vmatrix} 2p \\ \frac{1}{2} \\ 0 \end{vmatrix} = \begin{vmatrix} 0 & \bar{1} & 0 \\ 1 & 0 & 0 \\ 0 & 0 & 1 \end{vmatrix} + \begin{vmatrix} 0 \\ 0 \\ 0 \end{vmatrix} \tag{5.11}
$$

$$\mathbf{R}_m \qquad \mathbf{t}_m \qquad \mathbf{R}_b \qquad \mathbf{t}_b \qquad \mathbf{R}_4 \qquad \mathbf{t}_4$$

whence $q = 1/2$ and $p = q/2 = 1/4$, as before (see 5.4.9).

We can see that the difference between point groups and space groups is brought out by non-zero values of the translation vectors, $\mathbf{t}$. It may be found instructive to review the various techniques used to determine the relative positions of the symmetry elements of the space groups that we have studied (and others) in terms of the matrix method.

5.4.12 Symmetry symbols in three dimensions

The symmetry symbols in three-dimensional space groups are collected together, for reference purposes, in the next two tables.

Table 5.7—Notation for symmetry planes in three dimensions

Symbol	Graphic symbol		Glide translaton
m		⊥ paper ∥ paper	
a		⊥ paper ∥ paper	$a/2$ $a/2$
b		⊥ paper ∥ paper	$b/2$ $b/2$
c		⊥ paper	$c/2$
n		⊥ paper ∥ paper	$(a+b)/2$ or $(b+c)/2$ or $(c+a)/2$; $(a+b+c)/2$ in tetragonal and cubic
d		⊥ paper ∥ paper	$(a \pm b)/4$ or $(b \pm c)/4$ or $(c \pm a)/4$; $(a \pm b \pm c)/4$ in tetragonal and cubic†

† The arrow shows the direction of the horizontal components of translation when the z component is positive. In the symbols for any plane parallel to the paper, a fraction indicates the z height of that plane.

Table 5.8—Notation for symmetry axes in three dimensions

Symbol	Graphic symbol	Right-handed translation
1		
$\bar{1}$	○	
2	● ⊥ paper	
	→ ∥ paper	
2_1	⌡ ⊥ paper	$c/2$
	⟶ paper	$a/2$ or b/c
3	▲	
3_1	▲	$c/3$
3_2	▲	$2c/3$
$\bar{3}$	△	
4	◆	
4_1	◆	$c/4$
4_2	◆	$2c/4 = c/2$
4_3	◆	$3c/4$
$\bar{4}$	◈	
6	⬣	
6_1	⬣	$c/6$
6_2	⬣	$2c/6 = c/3$
6_3	⬣	$3c/6 = c/2$
6_4	⬣	$4c/6 = 2c/3$
6_5	⬣	$5c/6$
$\bar{6}$	⬠	

PROBLEMS 5

1. Draw a diagram to show the symmetry elements and general equivalent positions in plane group $c2mm$. List the coordinates of the general equivalent positions, and of the special equivalent positions in their correct sets.

2. Draw a diagram of the symmetry elements in plane group $p2mg$; take care not to put the two-fold rotation point at the intersection of m and g (Why?). On the diagram, insert each of the motifs P, V and Z in turn, using the minimum number of motifs consistent with the plane-group symmetry.

3. Draw a diagram to show the symmetry elements and general equivalent positions in $p3m1$. Write down the coordinates of the general positions.

4. (a) Complete the study of space group $P2_1/c$ (see 5.4.4) by giving the coordinates of the general and special equivalent positions, and the symmetries of the principal projections.

   (b) Biphenyl ⟨○⟩—⟨○⟩ crystallizes in space group $P2_1/c$ with two molecules per unit cell. What can be deduced about both the positions of the molecules in the unit cell, and the conformation of the molecule?

5. Show that space groups Pa, Pc and Pn represent the same pattern of symmetry elements.

6. What space group, in class 222, is obtained if the 2_1 axes parallel to x and z, in a P unit cell, intersect?

7. Draw a digram for space group Pmc ($mm2$ class). What is the nature of the symmetry axis along x?

8. Determine analytically the orientation of the symmetry elements in space group *Pban* (origin on $\bar{1}$). Confirm your results by drawing a space group diagram, and by the half-shift rule. What is the full symbol for this space group?

9. Draw a diagram for space group *Pbam*, with the origin at the intersection of the three symmetry planes. List the coordinates of the general positions. Transform the origin to a centre of symmetry, and deduce the coordinates of the general positions with respect to the new origin.

10. What do the standard symbols *Pban* and *Cmca* become if referred to the $\bar{c}ba$ setting?

11. Draw a diagram to show the symmetry elements and general positions in space group *P4nc*. List the general equivalent positions.

12. Repeat Problem 11, but for space group *P31c*.

13. The matrices for an *n*-glide plane operator normal to *a* and for an *a*-glde plane operator normal to *b* are

$$
\begin{vmatrix} \bar{1} & 0 & 0 \\ 0 & 1 & 0 \\ 0 & 0 & 1 \end{vmatrix} + \begin{vmatrix} 0 \\ \frac{1}{2} \\ \frac{1}{2} \end{vmatrix} \text{ and } \begin{vmatrix} 1 & 0 & 0 \\ 0 & \bar{1} & 0 \\ 0 & 0 & 1 \end{vmatrix} + \begin{vmatrix} \frac{1}{2} \\ 0 \\ 0 \end{vmatrix}
$$

Determine the nature and orientation of the symmetry operator arising from the combination of the two operators given.

14. Draw four points to form a parallelogram for which $a = b$ and $\gamma = 105°$. Extend the array to about four rows, with six points in each row. (a) On the left-hand side of the diagram, set a letter Z, with the mid-point of the oblique stroke of the letter, at each lattice point, with all letters in the same orientation. What is the plane group of the Z arrangement? (b) On the right-hand side of the diagram, use a letter H in place of Z, with the centre of the horizontal stroke of the letter at each lattice point, always in the same orientation. What is the plane group?

6

X-ray Diffraction

> Their (X-rays) moral is this—that a
> right way of looking at things will
> see through almost anything.
>
> Samuel Butler, 1912: *Notebooks*, Vol. V

6.1 INTRODUCTION

In so far as much of our detailed knowledge about the symmetry of crystals and molecules arises through X-ray studies, it becomes relevant to consider some aspects of X-ray diffraction at this stage.

6.2 X-RAYS

X-rays are electromagnetic radiation of short wavelength, and are produced by the sudden arrest of rapidly moving electrons at the surface of a target material. If an electron falls through a potential difference of V volt, it gains energy of Ve electronvolt, where e is the charge on an electron. If this energy were wholly converted into a quantum of X-ray photons, the photon wavelength would be given by

$$\lambda = hc/Ve \qquad (6.1)$$

where h and c are, respectively, the Planck constant and the speed of light *in vacuo*. A large proportion of the energy of the electrons is dissipated as heat in the target material, and the production of X-rays is not a very efficient process.

6.3 X-RAY SPECTRUM

The output of an X-ray tube consists of a range of wavelengths; a sharp minimum wavelength is given by (6.1), but the upper limit is indefinite. If the accelerating voltage is sufficiently large, electron transitions occur in the target material, and are accompanied by the emission of high intensity characteristic X-rays. In the case of a copper target, the characteristic spectrum consists of $K\alpha$ ($\bar{\lambda} = 0.1542$ nm) and $K\beta$ ($\lambda = 0.1392$ nm) lines together.

6.4 FILTERING

The continuous wavelength (white) radiation is always present, whether or no the characteristic spectrum is superimposed on it. An effectively monochromatic source can be produced by using a β-filter. For copper radiation, nickel foil (0.018 mm thickness) acts as a β-filter on account of its high resonance absorption (absorption edge) at a wavelength of 0.1487 nm. The intensities of all wavelengths are reduced by filtering because of the absorption law:

$$I = I_0 \exp(-\mu t) \tag{6.2}$$

I and I_0 are, respectively, the transmitted and incident intensities, t is the thickness of the material and μ is its linear absorption coefficient. Because the β line is preferentially absorbed, the high intensity $K\alpha$ line is dominant. In fact, $K\alpha$ is a doublet, $K\alpha_1$ and $K\alpha_2$; the wavelength $\bar{\lambda} = 0.1542$ nm is a weighted value.

6.5 X-RAY DIFFRACTION

The distances between atoms are commensurate with the wavelength of X-rays, and so a crystal acts as a three-dimensional diffraction grating for X-rays. X-ray diffraction is really a combined scattering and interference process. All groups of atoms scatter X-rays, but strong directional effects arise from the co-operative scattering of an immense number of unit cells that are packed together according to the principles that we have described.

For example, the tetrahedral atom groups SiO_4 may be regularly arranged in space according to a given pattern, as in α-quartz (Fig. 6.1) space group $P3_121$, and so

Fig. 6.1—Unit cell of α-quartz (SiO_2).

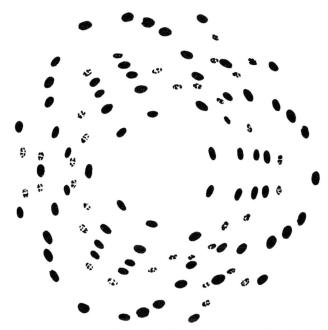

Fig. 6.2—Laue photograph taken with the X-ray beam along a three-fold axis; Laue symmetry 3*m*.

Fig. 6.3—Silica glass; the same SiO$_4$ units as in α-quartz, but arranged in a random manner.

give rise to a strong diffraction effect, characterized by a spot pattern such as Fig. 6.2 from a crystal of this substance. Alternatively, the SiO_4 units may be arranged in a random manner (Fig. 6.3), as in silica glass, and give rise to only a diffuse diffraction effect (Fig. 6.4).

To understand more about the important X-ray information, we must consider the conditions for X-ray diffraction from single crystals.

6.5.1 Bragg's equation

The diffraction of X-rays by crystals can be treated by a formal physical method, leading to the Laue equations for diffraction. An alternative method, due to Bragg, considers diffraction as reflexion of X-rays from crystal planes, an approach suggested by the results of early X-ray diffraction experiments.

Consider two adjacent planes in a family of parallel, equidistant planes (see Fig. 4.29) of spacing d, and let a parallel beam of X-rays be incident at a glancing angle θ (Fig. 6.5). The reflected beams also make an angle θ with the crystal planes. The

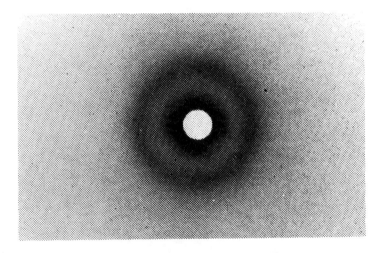

Fig. 6.4—Diffuse X-ray pattern from silica glass.

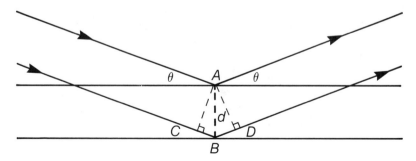

Fig. 6.5—Bragg equation: two planes of any family of parallel, equidistant planes are shown, with two typical X-ray paths indicated.

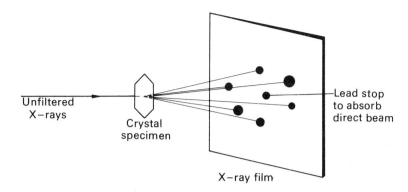

Fig. 6.6—Laue method, using a flat film.

X-rays are in phase along the normal AC to the incident wavefront, and also along AD normal to the reflected wavefront. The path difference for reflexion of the two typical rays from the planes is $BC + BD$. Since each of these lengths is equal to $d \sin \theta$, the total path difference is $2d \sin \theta$. For a reinforcement condition, we have

$$2d \sin \theta = n\lambda \tag{6.3}$$

where n is an integer; (6.3) is known as Bragg's equation.

Following Section 4.6, each possible family of planes in a lattice is characterized by its own interplanar spacing, so that

$$d_{nh,nk,nl} = d_{hkl}/n \tag{6.4}$$

Hence, the Bragg equation is generally given as

$$2d_{hkl} \sin \theta_{hkl} = \lambda \tag{6.5}$$

If this equation is satisfied for planes 1 and 2 in the family, it will be satisfied also for planes 2 and 3, 3 and 4, and so on for the whole family (hkl).

6.5.2 Laue photographs

In the simplest technique, the Laue method, a beam of white X-rays passes through a stationary single crystal, and the diffraction effects are recorded on a flat film (Fig. 6.6). Fig. 6.2 shows a typical Laue photograph.

The variables in (6.5) are θ, d and λ. For a given family of planes, d is fixed. If the crystal is stationary, θ is fixed too. The equation is satisfied through λ being a variable, the crystal actually selecting from the white radiation the wavelength appropriate for the Bragg condition to be fulfilled. It is in this condition that the distinction between Bragg reflexion and true optical reflexion is shown. In optical reflexion, as long as the incident ray meets the reflecting surface a reflexion is obtained. In the Bragg reflexion of X-rays, (6.5) must be satisfied, as well the normal reflexion geometry.

Each spot on an X-ray film (Fig. 6.2, for example) is characterized by two important parameters: a positional parameter, θ or the Miller indices hkl, dependent upon the crystal geometry, and an intensity parameter, governed by the nature and arrangement of the atoms in the crystal. In the Laue method, it is a practical difficulty with an unknown material that the particular wavelength producing a given spot is unknown. In addition, it is difficult to assign the positional parameters, except in simple cases.

We shall consider next satisfying the Bragg equation by varying θ in monochromatic conditions, but first we shall discuss the Ewald construction.

6.5.3 Ewald's construction

The Bragg construction requires us to think about a crystal as numerous families of crystal planes, which is not easy. The geometric interpretation of X-ray diffraction is facilitated greatly by a construction due to Ewald.

A sphere of radius one reciprocal space unit ($K = 1$ in (4.35)), imagined centred on the crystal C, is constructed on the X-ray beam as a diameter AQ (Fig. 6.7). From

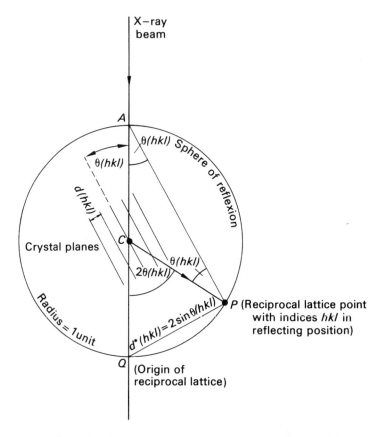

Fig. 6.7—Ewald's construction: an X-ray reflexion arises when a reciprocal lattice point P intersects the surface of the Ewald sphere; the direction of the reflected ray is CP. In a particular case of rotation about the x axis, normal to the circular section shown, this circle is the central or zero-layer circle, and all reflexions such as CP will be indexed as $0kl$.

the construction,

$$AQ = 2 \tag{6.6}$$

$$\widehat{APQ} = 90° \tag{6.7}$$

$$QP = AQ \sin \theta_{hkl} = 2 \sin \theta_{hkl} \tag{6.8}$$

Using (6.5)

$$QP = \lambda/d_{hkl} \tag{6.9}$$

which, from (4.35) with $K = \lambda$, is d^*_{hkl}. Thus, P is the reciprocal lattice point hkl, and CP is the direction of the hkl reflexion.

The occurrence and direction of X-ray reflexions can now be explained through the reciprocal lattice and the Ewald sphere (Fig. 6.8). The origin of the reciprocal lattice is at O, and the crystal is at C. As the crystal rotates about an axis through C, the reciprocal lattice rotates in a synchronous manner about a parallel axis through O. As a reciprocal lattice point passes through the Ewald sphere a reflexion occurs along the direction from C to that point.

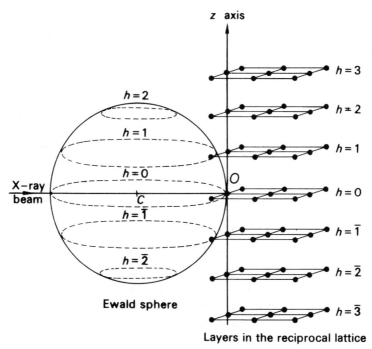

Fig. 6.8—Formation of layer lines in terms of Ewald's construction. If a cylindrical film encloses the crystal, such that the crystal rotation axis and the film axis coincide, layer lines are formed when the film is opened out (see Fig. 6.9).

If we imagine a cylindrical film placed about the crystal such that the film axis and the rotation axis coincide, then the X-ray reflexions intersect the film in circles of spots, which become straight lines (layer lines) when the film is opened out (Fig. 6.9). While this type of photograph contains positional information that can be retrieved, it would be more convenient if the photograph, which is a representation

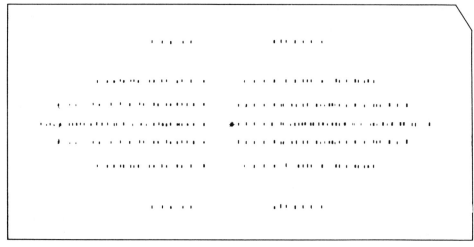

Fig. 6.9—Layer lines: for rotation about the x axis, the central line of spots (zero layer) is indexed 0kl.

of the reciprocal lattice, would present this information in an undistorted manner, layer by layer.

This result can be achieved by the precession method, in which a crystal axis t precesses about the X-ray beam, with the film always perpendicular to that crystal axis (Fig. 6.10). The resulting X-ray photograph is undistorted, a sort of *contact print* of the reciprocal lattice layer in question (Fig. 6.11). The positional parameters can be assigned to the spots by inspection. Thus, we can see from the photograph that reflexions occur systematically only for $k + l = 2n$, which indicates an n-glide plane normal to the x axis (see 6.7.1).

6.6 INTENSITY EXPRESSIONS

The intensities of the X-ray reflexions may be given in the form

$$I_{hkl} = Ap\xi LK|F_{hkl}|^2 \tag{6.10}$$

The parameters A, p, ξ and L are connected with absorption, polarization, extinction and the geometry of the experiment; K is a scale factor, and $|F_{hkl}|$ is the modulus of the structure factor. For our purposes, we are concerned with F_{hkl} only; $|F_{hkl}|^2$ may be called the ideal intensity, the intensity in the absence of the various correction factors in (6.10).

6.6.1 Structure factor

The structure factor F_{hkl} is given by the equation

$$F_{hkl} = \sum_{j=1}^{N} f_{j,\theta} \exp[i2\pi(hx_j + ky_j + lz_j)] \tag{6.11}$$

F_{hkl} is the structure factor for the family of hkl planes, and the summation is over all N atoms in the unit cell; $f_{j,\theta}$ is the scattering factor of the j atom, and its value

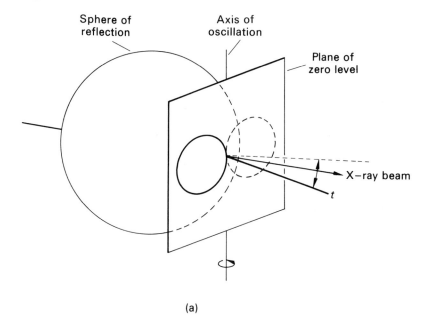

(a)

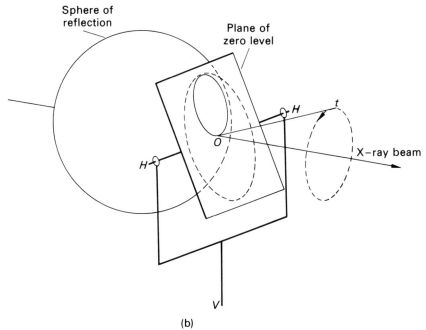

(b)

Fig. 6.10—Rotation and precession geometry compared: (a) reciprocal lattice zero layer with its normal *t* oscillating (restricted rotation) by a given amount on each side of the X-ray beam. The maximum symmetry possible is that of the oscillation movement, *mm*2, (b) reciprocal lattice zero layer with its normal *t* precessing about the X-ray beam (*H* and *V* are horizontal and vertical axes). The film is tangential to the Ewald sphere at the position corresponding to an emergent X-ray reflexion, so that the reciprocal lattice is recorded in an undistorted manner. An undistorted photograph could also be produced by using a spherical film, with the crystal at the centre of the sphere, but this method presents special practical difficulties.

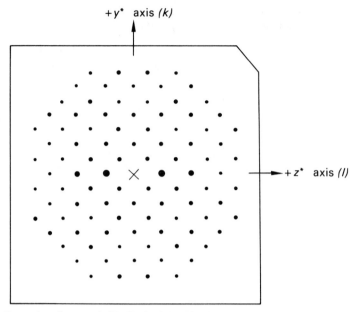

Fig. 6.11—Precession photograph (idealized) of the 0*kl* reciprocal lattice layer of an orthorhombic crystal. The spots can be in indexed by inspection, and we can see that thre are no reflexions for $k + l = 2n + 1$. Thus, we identify, from this information alone, an *n* glide plane normal to the *x* axis.

depends upon the Bragg angle θ for a given wavelength of X-rays; x_j, y_j, z_j are the fractional coordinates of the jth atom. The equation expresses the combined scattering of all atoms in the unit cell; at any finite temperature, the $f_{j,\theta}$ terms would need modifying to allow for thermal vibrations, but we shall not be concerned with this feature.

The structure factor is a complex quantity, and can be separated into its real (A') and imaginary (B') components:

$$F_{hkl} = A'_{hkl} + iB'_{hkl} \tag{6.12}$$

where

$$A'_{hkl} = \sum_{j=1}^{N} f_{j,\theta} \cos 2\pi(hx_j + ky_j + lz_j) \tag{6.13}$$

and

$$B'_{hkl} = \sum_{j=1}^{N} f_{j,\theta} \sin 2\pi(hx_j + ky_j + lz_j) \tag{6.14}$$

The conjugate term $F_{\bar{h}\bar{k}\bar{l}}$ is given by

$$F_{\bar{h}\bar{k}\bar{l}} = A'_{\bar{h}\bar{k}\bar{l}} + iB'_{\bar{h}\bar{k}\bar{l}} \tag{6.15}$$

It follows that

$$|F_{hkl}|^2 = |F_{\bar{h}\bar{k}\bar{l}}|^2 \tag{6.16}$$

which is Friedel's law; it means that the diffraction pattern of a crystal will be centrosymmetric, whatever the symmetry of the crystal. Thus, crystals of classes 2, *m* and 2/*m* will, normally, behave towards X-rays as though they all belonged to 2/*m*. Friedel's law breaks down seriously when any atoms in the crystal are irradiated with X-rays of wavelength near to an absorption edge, and less seriously with heavy atoms and various radiations. Our ensuing discussions will be confined to situations within Friedel's law.

6.6.2 Laue projection symmetry

The centrosymmetric point groups (Laue groups) have been discussed in Section 3.6.5, and projected symmetry has been treated in Section 3.6.6. In considering projected point-group symmetry, we carried out the projection and then asked about the symmetry of the two-dimensional projection (Table 3.9). In Laue projection symmetry, we consider what the X-ray beam 'sees' as it passes through the crystal, and it is this information that is projected on to the film by the X-ray beam (Fig. 6.2). To obtain the required result for, say an X-ray beam travelling normal to {110} in a crystal of point group 422, we consider the corresponding Laue group, $\frac{4}{m}mm$ (Fig. 3.18(d)). The X-ray beam encounters two-fold symmetry along its direction of travel, and lies in two *m* planes. Thus, the Laue projection symmetry is 2*mm*. Table 6.1 lists the Laue projection symmetry for the thirty-two crystal classes.

Table 6.1—Laue projection symmetry

System	Point groups	Laue groups	Point symmetry		
			{100}	{010}	{001}
Triclinic	1, $\bar{1}$	$\bar{1}$	1	1	1
Monoclinic	2, *m*, 2/*m*	2/*m*	*m*	2	*m*
Orthorhombic	222, *mm*2, *mmm*	*mmm*	2*mm*	2*mm*	2*mm*
			{001}	{100}	{110}
Tetragonal	4, $\bar{4}$, 4/*m*	4/*m*	4	*m*	*m*
	4*mm*, 422,	$\frac{4}{m}mm$	4*mm*	2*mm*	2*mm*
	$\bar{4}2m$, $\frac{4}{m}mm$				
			{0001}	{10$\bar{1}$0}	{11$\bar{2}$0}
Hexagonal	6, $\bar{6}$, 6/*m*	6/*m*	6	*m*	*m*
and Trigonal	6*mm*, 622	$\frac{6}{m}mm$	6*mm*	2*mm*	2*mm*
(hexagonal axes)	$\bar{6}m2$, $\frac{6}{m}mm$				
	3, $\bar{3}$	$\bar{3}$	3	1	1
	3*m*, 32, $\bar{3}m$	$\bar{3}m$	3*m*	*m*	2
			{100}	{111}	{110}
Cubic	23, *m*3	*m*3	2*mm*	3	*m*
	432, $\bar{4}3m$, *m*3*m*	*m*3*m*	4*mm*	3*m*	2*mm*

6.7 Applications of the structure factor equation

The reciprocal lattice points may be weighted in proportion to the value of $|F|^2$ for the various reflexions. This weighted reciprocal lattice contains the basic structural and geometrical information obtained by X-ray diffraction. The weighted reciprocal lattice can show different intensity distributions. One of them, the distribution according to Laue symmetry, has been discussed. Another important distribution, also governed by the crystal geometry, relates to systematically absent reflexions.

6.7.1 Systematic absences—limiting conditions

Certain groups of possible X-ray reflexions may be absent on account of the crystal geometry, irrespective of the nature and positions of the atoms in the unit cell. Systematic absences arise when there is centring present, either generally, or in a particular zone or just along a row in reciprocal space. We will consider an example of each of these three types.

Centred unit cell
Consider a unit cell centred on the C faces. This description implies that the atoms in the unit cell are related in pairs as x, y, z and $\frac{1}{2} + x, \frac{1}{2} + y, z$. From (6.11),

$$F_{hkl} = \sum_{j=1}^{N/2} f_{j,\theta} \left\{ \exp[i2\pi(hx_j + ky_j + lz_j)] + \exp\left[i2\pi(hx_j + ky_j + lz_j + \frac{h+k}{2}) \right] \right\} \quad (6.17)$$

which may be written as

$$F_{hkl} = \sum_{j=1}^{N/2} f_{j,\theta} \exp[i2\pi(hx_j + ky_j + lz_j)] \left[1 + \exp\left(i2\pi \frac{h+k}{2} \right) \right] \quad (6.18)$$

Since h and k are integers the term on the extreme right-hand side of (6.18) is either 2 $(h + k = 2n)$ or zero $(h + k = 2n + 1)$, where n is an integer. Thus, the reflexions hkl are systematically absent when the sum of h and k is an odd integer. Expressed another way, the hkl reflexions are present for the sum of h and k equal to an even integer. In the latter form, they are the conditions limiting X-ray reflexions, or just *limiting conditions*, and are listed, where appropriate, as the extreme right-hand column of the space group descriptions (see Fig. 5.12), for example). Each unit cell type leads to certain limiting conditions (Table 6.2).

Table 6.2—Limiting conditions for unit cell types

Unit cell type	Limiting condition
P	hkl: None
A	hkl: $k + l = 2n$
B	hkl: $h + l = 2n$
C	hkl: $h + k = 2n$
I	hkl: $h + k + l = 2n$
F	hkl: $h + k = 2n,\ k + l = 2n,\ (l + h = 2n)$†
R	hkl: None
R_{hex} (obverse)	hkl: $-h + k + l = 3n$

† This condition is not independent of the other two conditions.

Glide plane symmetry

Consider a c glide plane normal to the y axis; it relates atoms in pairs as x, y, z and $x, \bar{y}, \frac{1}{2} + z$. From (6.11), we have

$$F_{hkl} = \sum_{j=1}^{N/2} f_{j,\theta} \{ \exp[i2\pi(hx_j + ky_j + lz_j)] + \exp[i2\pi(hx_j - ky_j + lz_j + l/2)] \} \quad (6.19)$$

or

$$F_{hkl} = \sum_{j=1}^{N/2} f_{j,\theta} \exp[i2\pi(hx_j + lz_j)][\exp(i2\pi ky_j) + \exp(i2\pi(-ky_j + l/2)] \quad (6.20)$$

To proceed further, we must set k equal to zero, thus restricting our analysis to the $h0l$ zone of reflexions. Thus,

$$F_{h0l} = \sum_{j=1}^{N/2} f_{j,\theta} \exp[i2\pi(hx_j + lz_j)][1 + \exp(i2\pi l/2)] \quad (6.21)$$

Now we see, from the form the term on the extreme right-hand side of (6.21), that F_{h0l} will be zero if l is an odd integer. Thus, the limiting condition for a c glide plane normal to the y axis is $h0l: l = 2n$, where n is an integer.

Screw axis

For a 2_1 screw axis along $[0, y, 0]$, the atoms in the unit cell are related in pairs as x, y, z and $\bar{x}, \frac{1}{2} + y, \bar{z}$. From (6.11) and a similar analysis as in the previous cases, we find

$$F_{hkl} = \sum_{j=1}^{N/2} f_{j,\theta} \exp(i2\pi ky_j)[\exp(i2\pi(hx_j + lz_j)) + \exp(i2\pi(-hx_j - lz_j + k/2))] \quad (6.22)$$

To proceed, we set $h = l = 0$, whence

$$F_{0k0} = \sum_{j=1}^{N/2} f_{j,\theta} \exp(i2\pi ky_j)[1 + \exp(i2\pi k/2)] \quad (6.23)$$

Hence, F_{0k0} will be zero for $k = 2n + 1$, where n is an integer, so the limiting condition is $0k0: k = 2n$. Similar results would be obtained for any *position* of the given translational symmetry elements in the given *orientation*. Tables 6.3 and 6.4 list the limiting conditions associated with glide planes and screw axes.

Table 6.3—Limiting conditions for glide planes

Reflexion	Orientation	Limiting condition	Glide plane	System
$hk0$	(001)	$h = 2n$ $k = 2n$ $h + k = 2n$ $h + k = 4n$	a b n d	Tetragonal / Orthorhombic
$0kl$	{100}	$k = 2n$ $l = 2n$ $k + l = 2n$ $k + l = 4n$	b c n d	Orthorhombic Tetragonal Cubic
$h0l$	(010)	$h = 2n$ $l = 2n$ $h + l = 2n$ $h + l = 4n$	a c n d	Monoclinic† / Orthorhombic
$hh2\bar{h}l$ $hh\bar{h}0l$ hhl	{1$\bar{1}$00} {11$\bar{2}$0} {110}	$l = 2n$ $l = 2n$ $l = 2n$ $2h + l = 4n$	c c c d	Hexagonal Hexagonal Rhomdohedral) Tetragonal) Cubic

† The y axis is unique in the monoclinic system.

Table 6.4—Limiting conditions for screw axes

Reflexion	Orientation	Limiting condition	Screw axis	System
$h00$	$\langle 100 \rangle$	$h = 2n$	2_1 4_2	Orthorhombic ⎫
		$h = 4n$	$4_1, 4_3$	Tetragonal ⎬ Cubic ⎭
$0k0$	$[010]$	$k = 2n$	2_1	Monoclinic† Orthorhombic
$00l$	$[001]$	$l = 2n$	2_1	Orthorhombic
			4_2 ⎫	Tetragonal
		$l = 4n$	$4_1, 4_3$ ⎭	
$000l$	$[0001]$	$l = 2n$ $l = 3n$	6_3 $3_1, 3_2$ ⎫	Hexagonal
		$l = 6n$	$6_2, 6_4$ $6_1, 6_5$ ⎭	

† The y axis is unique in the monoclinic system.

6.8 USING THE DIFFRACTION INFORMATION

If we have obtained a diffraction record for a crystal and have indexed (assigned the *hkl* indices to) the reflexions, we are in a position to look for systematic conditions among the data that may lead to space group determination. We shall consider three examples here, and others in the Problems section.

Example 1	*Conclusions*
Laue group $2/m$	Monoclinic, point group 2, m or $2/m$
hkl: None	P unit cell
$h0l: l = 2n$	c glide plane normal to the y axis
$0k0$: None	Space group Pc or $P2/c$

Example 2
$hkl: h + k = 2n$ C unit cell
$0kl: (k = 2n)$
$h0l: l = 2n, (h = 2n)$ c glide plane normal to the y axis
$hk0: (h + k = 2n)$
$h00: (h = 2n)$
$0k0: (k = 2n)$
$00l: (l = 2n)$

It is important to work down the hierarchy of reflexions, general (*hkl*), special (one zero index), special (two zero indices), noting conditions that are dependent upon more general results. Thus, in Example 2, if $h + k = 2n$ for *hkl*, then it is to be expected that we find $0kl: k = 2n$, and so on; these non-independent conditions are placed in parentheses. For *h0l*, $h = 2n$ derives from the first-listed condition, but $l = 2n$ for the same reflexions is independent, and corresponds to a c glide plane normal to the y axis. We conclude that the space group is either Cc or $C2/c$. Often diffraction data alone lead to a choice of possible space group. Only 50 out of the 230 space groups are uniquely determined by the diffraction data. Generally, the distinction rests upon whether or no there is a centre of symmetry present. This addition datum leads to 192 of the 230 space groups being uniquely identified.

Fortunately, statistical tests can be applied to the whole intensity distribution, or to certain zones or rows of reflexions. A full discussion of this topic lies outside the scope of this book. However, we may note that in the case of Example 2, the statistics of the distribution of $|F_{hkl}|^2$ values would show an acentric (A) distribution for space group Cc, but a centric (C) distribution for $C2/c$, thus resolving the problem of space group determination.

In another possible case, we may find only evidence for monoclinic C: hence, the space group could be $C2$, Cm or $C2/m$. In such a case, we can use the statistical tests in conjunction with Table 3.9 (the projected symmetry). Thus, we have:

Space group	Distribution		
	hkl	$h0l$	$0k0$
$C2$	A	C	A
Cm	A	A	A
$C2/m$	C	C	C

Example 3

An orthorhombic crystal produced the following X-ray data. In addition, the distribution of intensities in the $0kl$ zone was found to be acentric. What is the space group?

hkl	hkl	hkl	hkl
111	011	210	020
112	015	220	060
212	024	420	080
312	031	430	002
322	203	200	004
332	303	400	008

Summarizing the results, we have

hkl: None—P unit cell

$0kl$: $k + l = 2n$—n glide plane normal to the x axis

$h0l$: None

$hk0$: $h = 2n$—a glide plane normal to the z axis

$h00$: $(h = 2n)$

$0k0$: $(k = 2n)$

$00l$: $(l = 2n)$

From the diffraction results, the space group is either $Pnma$, or $Pn2_1a$ which is the $a\bar{c}b$ setting of $Pca2_1$. If we now add in the statistical information (see Table 3.9), $Pnma$ would be excluded and the correct result is $Pn2_1a$. We know that the axis in the symbol is 2_1 and not 2, because of the $c/2$ translation that arises from the n glide.

We cannot show the presence of 2_1 from the condition $0k0: k = 2n$, because this condition is not independent of the condition for the n glide symmetry.

We have dealt with X-ray matters briefly here, and for further detail the reader should consult one of the books mentioned in the Bibliography. Nevertheless, the material treated here should prove to be sufficient for the present purposes of understanding the derivation and meaning of limiting conditions, or systematic absences, and using them, together with other data, to determine the space group of a crystal.

PROBLEMS 6

1. An X-ray tube is operated at 30 kV. What is the minimum wavelength of X-rays obtainable from such a tube?
2. An X-ray beam passes through a nickel foil of thickness 0.020 mm. If the linear absorption coefficient for nickel is $44 \, \text{mm}^{-1}$, what fraction of incident intensity would be transmitted through the foil?
3. The unit cell dimension for potassium chloride is 0.628 nm. What is the value of d_{123}, and at what value of the Bragg angle θ would reflexion occur from the (123) family of planes, using copper $K\alpha$ radiation ($\bar{\lambda} = 0.1542$ nm)?
4. Fig. 6.1 is a Laue photograph recorded on a flat-plate film placed normal to a beam of white X-rays. What is the Laue symmetry shown by the film? Given that the crystal is cubic, what direction in the crystal is parallel to the X-ray beam?

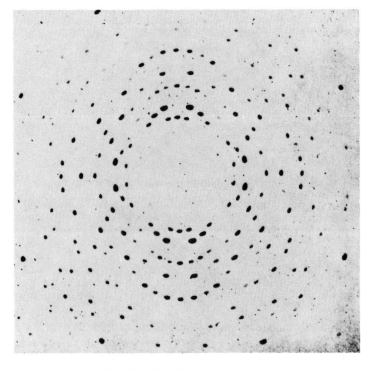

Fig. P6.1—Laue flat-film photograph

5. From the structure factor equation, determine the limiting conditions for X-ray reflexion from a body-centred unit cell.

6. In a crystal, the atoms present in the unit cell are related by a two-fold screw axis along $[\frac{1}{4}, 0, z]$. Deduce what reflexions would be systematically absent on account of this symmetry?

7. Determine the limiting conditions for an n glide plane normal to the z axis.

8. What space groups are indicated by the following data?

Monoclinic (y unique)
(a) hkl: None
    $h0l$: None
    $0k0$: $k = 2n$
(b) hkl: $h + k = 2n$
    $h0l$: $(h = 2n)$
    $0k0$: $(k = 2n)$

Orthorhombic
(c) hkl: None
    $0kl$: $k = 2n$
    $h0l$: None
    $hk0$: None
    $h00$: None
    $0k0$: $(k = 2n)$
    $00l$: None
(d) hkl: $h + k + l = 2n$
    $0kl$: $k = 2n, l = 2n$
    $h0l$: $l = 2n, h = 2n$
    $hk0$: $h = 2n, k = 2n$
    $h00$: $h = 2n$
    $0k0$: $k = 2n$
    $00l$: $l = 2n$
(The non-independent conditions have not been parenthesized.)

Tetragonal
(e) hkl: $h + k + l = 2n$
    $hk0$: h (or k) $= 2n$
    $00l$: $l = 4n$
(f) hkl: None
    $0kl$: $l = 2n$
    hhl: $l = 2n$

Laue group $\dfrac{4}{m} mm$

(g) In (a), how would the deduction be affected if the crystal contained two molecules of a steroid molecule per unit cell?

(h) Rewrite the conditions in (d), putting the non-independent conditions in parentheses.

9. Write the structure factor equation for a crystal with a centre of symmetry at the origin, and simplify it as far as possible.

10. The structure factor equation (6.11) can be written as

$$F_{hkl} = \sum_{j=1}^{N/R} f_{j,\theta} \sum_{m=1}^{R} \exp[i2\pi(hx_m + ky_m + lz_m)$$

where the inner summation is over the number of general equivalent positions in the space group; it is often called the geometrical structure factor. Use the result from Problem 9 to establish the geometrical structure factor for space group $P2_1/c$. Hence, deduce the limiting conditions, and show that $F_{hkl} = (-1)^{k+l}F_{h\bar{k}l}$.

The identity $\cos A + \cos B = 2\cos \dfrac{A+B}{2} \cos \dfrac{A-B}{2}$ may be helpful.

7

Elements of Group Theory

The growth (of observation) consists
in a continual analysis of facts of
rough and general observation into
groups of facts more precise and minute

Walter Pater, 1889: *Appreciations*, 'Coleridge'

7.1 INTRODUCTION

By tradition, though not of necessity, group theory and its applications to chemistry is usually presented in the Schoenflies symmetry notation. We shall follow this convention though, here and there, some duplication will occur in the Hermann–Mauguin notation, so as to show that a single symmetry notation might be a worthy development. First, then, we shall consider the Schoenflies notation, and how it differs from that used so far in this book. Since we have done the groundwork in symmetry, the new notation should not be a problem.

7.2 SCHOENFLIES SYMMETRY NOTATION

Reference has been made to the Schoenflies notation in Section 1.4, and in Appendix 2 we compared the Schoenflies and Hermann–Mauguin notations for point groups. In addition, both notations were given in Figs. 3.11, 3.13, 3.17 to 3.21 and 3.27 to 3.34, and reference to these figures will be useful in the ensuing sections.

7.2.1 Rotation symmetry axes

The notation C_R is used ($R = 1, 2, 3, \ldots, \infty$), where C stands for *cyclic*. An operation C_R has the same meaning as before (see Section 3.5.1). Thus, C_2 is the symbol for a two-fold rotation axis, and so on; R is the degree of the *principal* symmetry axis.

7.2.2 Mirror reflexion symmetry plane

The notation σ is used (German: *spiegel* = mirror). The mirror plane that is normal to the principal C_R axis is called σ_h (horizontal), and mirror planes that contain the principal axis are designated σ_v (vertical). Thus $2/m$ is C_{2h}, whereas $mm2$ is C_{2v}; notice that in the point group symbols, σ is not used.

7.2.3 Alternating symmetry axis

This symmetry element is new to our discussion. It is symbolized by S_R, a single operation, and it takes a body from one state to an indistinguishable state for a rotation of $(360/R)°$ followed by reflexion through a plane normal to the line of the S_R axis. It is important to note that, here, the plane is not necessarily a mirror plane, although it acts like one in defining the S_R operation. Just as $\bar{R}$ is not always centrosymmetric, so S_R does not always include a horizontal $\sigma(m)$ plane. Thus, in S_3 there is a mirror plane normal to this axis, but in S_4 there is not. In particular, S_R should not be thought of as the product $\sigma_h \cdot C_R$, since this expression implies that S_R, σ_h and C_R are all present together.

The inversion axis ($\bar{R}$) is not used in the Schoenflies notation. Its place is taken by S_R, but S_R is not always equivalent to $\bar{R}$ for the same value of R: S_4 is equivalent to $\bar{4}$, but S_6 is the same as $\bar{3}$, and S_3 is equivalent to $\bar{6}$.

7.2.4 Point groups in the Schoenflies notation

By analogy with Table 3.8, we can write down a scheme summarizing the crystallographic point groups in the Schoenflies notation, Table 7.1. The D_R groups have R two-fold axes perpendicular to the principal C_R axis. If there are R vertical mirror planes bisecting the two-fold axes, the notation D_{Rd} is used. In the cubic point groups, because of the characteristic four three-fold rotation axes present in each group, the special symbols T (tetrahedral) and O (octahedral) are introduced. Symbols for some non-crystallographic point-groups are given in Fig. 3.34.

Table 7.1—Crystallographic point groups

	Triclinic	Monoclinic	Trigonal	Tetragonal	Hexagonal	Cubic
C_R	C_1	C_2	C_3	C_4	C_6	T
S_R	C_i (S_2)	C_s (S_1)	S_6	S_4	(S_3)†	
C_{Rh}		C_{2h}	C_{3h}†	C_{4h}	C_{6h}	T_h
		Orthorhombic				
D_R		D_2	D_3	D_4	D_6	O
C_{Rv}		C_{2v}	C_{3v}	C_{4v}	C_{6v}	
D_{Rd}			D_{3d}	D_{2d}		T_d
D_{Rh}		D_{2h}	D_{3h}	D_{4h}	D_{6h}	O_h

† C_{3h} is used in preference to S_3.

In the symbols C and D (German: *diedergruppe* = dihedral group) the Schoenflies notation emphasizes the highest principal proper rotation axis. Thus, C_{3h} ($\bar{6}$) and D_{3h} ($\bar{6}m2$) appear under the trigonal heading, although, crystallographically, they are hexagonal.

We shall now make more use of the identity symmetry element, and we give it the symbol I rather than E, to avoid confusion in a later context. It could be argued that I is the usual symbol for icosahedral symmetry, but that topic is not addressed herein.

7.3 GROUP THEORY

Group theory deals with sets of operations having the property that, when two operations of the set are carried out in succession, the result is that which could be attained by another, single operation in the set, starting from the same initial state (see also Section 3.6.4).

The combination of two operations will be called their *product*, although it may not always refer to a multiplication process. The product of two operations is an operation of the first kind (proper rotation) or second kind (improper rotation) according as the two operations are respectively of the same or of different kinds.

Axes of symmetry of the first or second kind, including the plane of symmetry and the centre of symmetry, are called symmetry operators; generally, they will have a specific orientation relative to a set of reference axes. When considered to be present in a body, we have referred to them as symmetry elements (Section 3.2).

7.3.1 Group postulates

A *group* consists of a set of mathematical objects called the *members* of the group, that satisfy the following conditions:

(a) The product of any two members of the set and the square of each member of the set are also members of the set.
(b) The associative law holds, that is, if a, b and c are three members of the set, then $a.b(c) = a(b.c)$.
(c) The set contains the identity member I, such that $a.I = I.a = a$ for each member a of the set.
(d) Each member of the set has an inverse that is also a member of the set, that is, for each member a, there exists another member a^{-1} in the set, such that $a.a^{-1} = a^{-1}.a = I$.

7.3.2 Group definitions

(a) A group may be *finite*, the number of members of the group being the *order* of the group, or *infinite*.
(b) An *abstract group* is concerned only with the relationships in the set of operations, there being no specific interpretation attached to the members of the group.
(c) A group in which all pairs of members commute, that is, where $a.b = b.a$, is called *Abelian*. Thus, in point group C_{2v}, $\sigma.C_2 = C_2.\sigma$ ($m.2 = 2.m$) so that C_{2v} is an Abelian group.
(d) A group consisting of a single member a and its powers a^2, a^3, ..., a^p, is called a *cyclic* group of order p, where p is the smallest integer for which $a^p = I$. Thus, C_4 is a cyclic group of order 4, contaning the members C_4, C_4^2 ($= C_2$), C_4^3 ($= C_4^{-1}$), C_4^4 ($= I$). We remarked in Section 3.2 that a given symmetry element may be associated with more than one operation, and that property is evident here.
(e) A sub-set of members of a group forming another group by itself is a sub-group of the original group. Thus, C_2 and C_s are sub-groups of C_{2v}; I is a sub-group of every group. (See also Section 3.4.6.)
(f) Every geometrical body, or crystal, or molecule is characterized by a set of symmetry operations that constitute a point group. A point group is a finite

group; its order is determined by the number of members of the group. The order is also the number of general equivalent points generated by the point group: you may recall that we suggested looking for a relationship between any one point on a diagram and all other points so as to ensure that all symmetry elements had been found and inserted (Section 5.3.1). Thus, we can produce a stereogram for the general equivalent points of D_{2h} (order 8), as in Fig. 3.17(c), given only the three mutually perpendicular mirror planes. However, when we ask how we reach all points on the diagram from any one of them by a single operation, we discover the three mutually perpendicular two-fold axes, normal to the mirror planes, the centre of symmetry and, trivially, the identity operation.

(g) Two groups G and G' are isomorphous if there is a one-to-one correspondence between the members of the groups.

(h) The properties of the members of a group are presented conveniently as a *group multiplication table*. The members of the group are listed in the top row, identity first, and again in the first column. The top row are considered as post factors and the first column as pre-factors. The entries in the table are the 'products' of the pre-factors and the post-factors.

7.3.3 Some examples in group theory

(a) Consider first an abstract group consisting of the members I, a and b, for which the relations $a^2 = b$ and $a.b = I$ hold. The group table is shown below.

	I	a	b
I	I	a	b
a	a	b	I
b	b	I	a

It follows from the relations given that $b^2 = a$; $a^2.b = b^2 = a.(ab) = a.I = I$.

A group consisting of three-fold rotations comprises the members I, C_3 and C_3^2. Instead of C_3^2 we could write C_3^{-1}, the inverse of C_3, because $C_3^{-1}.C_3 = C_3^2.C_3 = I$. The multiplication table becomes:

	I	C_3	C_3^2
I	I	C_3	C_3^2
C_3	C_3	C_3^2	I
C_3^2	C_3^2	I	C_3

Finally, consider the group of positive integers modulo 3, with *addition* as the law of combination. The group table is as follows:

	0	1	2
0	0	1	2
1	1	2	0
2	2	0	1

These three groups are isomorphous. There is a one-to-one correspondence between the members of the groups:

$$a \rightarrow C_3 \rightarrow 1$$
$$b \rightarrow C_3^2 \rightarrow 2$$
$$I \rightarrow C_3^3 \rightarrow 0$$

The groups are also Abelian; this nature can be detected by the symmetry of the entries across the principal diagonal (top left to bottom right) of the table.

(b) There is only one crystallographic point group of order unity, namely, C_1. The three groups of order two are C_i, C_s and C_2. The only crystallographic group of order three is the cyclic group C_3; all cyclic groups are Abelian.

(c) The point groups C_{3v} and D_3 are isomorphous, and can be given a common group table:

	I	C_3	C_3^{-1}	a	b	c
I	I	C_3	C_3^{-1}	a	b	c
C_3	C_3	C_3^{-1}	I	b	c	a
C_3^{-1}	C_3^{-1}	I	C_3	c	a	b
a	a	c	b	I	C_3^{-1}	C_3
b	b	a	c	C_3	I	C_3^{-1}
c	c	b	a	C_3^{-1}	C_3	I

In this table, the symbols a, b and c stand for the three mirror planes in C_{3v} and for the three two-fold axes in D_3. The groups are of interest as the lowest order non-Abelian groups; it may be noted that $a.b = C_3^{-1}$, where $b.a = C_3$. In the Hermann–Mauguin notation, the table for 32 (D_3) would be as follows:

	1	3	3^{-1}	2_x	2_y	2_u
1	1	3	3^{-1}	2_x	2_y	2_u
3	3	3^{-1}	1	2_u	2_x	2_y
3^{-1}	3^{-1}	1	3	2_y	2_u	2_x
2_x	2_x	2_y	2_u	1	3	3^{-1}
2_y	2_y	2_u	2_x	3	1	3^{-1}
2_u	2_u	2_x	2_y	3^{-1}	3	1

(d) An important group in crystallography is the translation group. The basic translations (repeat vectors) of a unit cell, $n_1\mathbf{a} + n_2\mathbf{b} + n_3\mathbf{c}$ (where the n_i are integral, including zero), form an infinite group under the group operation of vector addition; the zero vector stands for the identity operation.

(e) Space groups are infinite groups. Every space group of a given crystal class is isomorphous with a point group of that class (notice again the distinction between crystal class and point group). Consider point group D_2 (222): the general equivalent positions may be given the coordinates x, y, z; $\bar{x}, \bar{y}, z$; $\bar{x}, y, \bar{z}$; $x, \bar{y}, \bar{z}$. The equivalent positions for all space groups in this crystal class (see chart in Section 5.4.7) are as given above, modified by the presence of screw axes and by the displacements of the symmetry axes with respect to the space group origin. It was considerations such as this that led to the Hermann–Mauguin notation, which is the superior for space groups.

7.3.4 Notes on reference axes

The principal proper rotation axis is assigned to the z reference axis direction. Then the x and y axes are drawn mutually perpendicular. This convention applies to all symmetries, unlike with crystallographic axes.

In setting reference axes in a molecule, the z axis is chosen as above. If there are several equivalent symmetry axes, the z axis is that (principal) symmetry axis passing through the greatest number of atoms. If the molecule is planar and the z axis lies in the plane, the x axis is normal to the plane. If the molecule is planar and the z axis is perpendicular to the plane, the x axis lies in the plane, and passes through the greatest number of atoms. If the molecule is non-planar, the largest planar fragment is used to set the x axis, according to the above rules. In other cases, the selection of x is arbitrary, subject to the fact that it is perpendicular to z; the y axis is perpendicular to both the x and the z axes, and x, y and z form a right-handed set.

7.3.5. Similarity transformations

Two symmetry operations a and b are said to belong to the same *class* (not to be confused with crystal class) if there is some operation r, such that

$$r \cdot a \cdot r^{-1} = b \tag{7.1}$$

Then, b is said to be the *similarity transform* of a, and a and b are conjugate to each other. The operation σ is its own inverse. Hence, we can perform a similarity transformation on C_4 by finding the operation equivalent to $\sigma \cdot C_4 \cdot \sigma$. The result may be found graphically (Fig. 7.1), or by matrix multiplication.

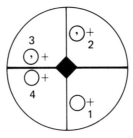

Fig. 7.1—Similarity transformation of C_4 on σ.

Let C_4 be along the z axis, with σ normal to the x axis. Then, we have:

$$
\begin{vmatrix} 0 & \bar{1} & 0 \\ 1 & 0 & 0 \\ 0 & 0 & 1 \end{vmatrix}
\cdot
\begin{vmatrix} \bar{1} & 0 & 0 \\ 0 & 1 & 0 \\ 0 & 0 & 1 \end{vmatrix}
=
\begin{vmatrix} 0 & \bar{1} & 0 \\ \bar{1} & 0 & 0 \\ 0 & 0 & 1 \end{vmatrix}
\tag{7.2}
$$

$$\qquad C_4 \qquad\qquad \sigma \qquad\qquad C_4 \cdot \sigma$$

$$\begin{vmatrix} \bar{1} & 0 & 0 \\ 0 & 1 & 0 \\ 0 & 0 & 1 \end{vmatrix} \cdot \begin{vmatrix} 0 & \bar{1} & 0 \\ \bar{1} & 0 & 0 \\ 0 & 0 & 1 \end{vmatrix} = \begin{vmatrix} 0 & 1 & 0 \\ \bar{1} & 0 & 0 \\ 0 & 0 & 1 \end{vmatrix} \qquad (7.3)$$

$$\sigma \qquad\qquad C_4 . \sigma \qquad\qquad \sigma . C_4 \sigma = C_4^{-1}$$

Thus, C_4 and C_4^{-1} $(=C_4^3)$ are in the same class. In the same way, we can show that

$$C_4 . \sigma . C_4^{-1} = \sigma' \qquad (7.4)$$

where σ' is normal to the y axis. Thus, σ and σ' are in the same class. This result should not surprise us, as σ and σ' are equivalent under the four-fold rotation. What about C_4^2? We see that

$$C_4 . C_4^2 . C_4^{-1} = C_4^2 = C_4^2 \qquad (7.5)$$

so that C_4^2 is its own inverse, which is to be expected because $C_4^2 = C_2$, but C_4^2 is not of the same class as C_4 and C_4^{-1}; it is often referred to as C_2.

In C_{4v}, there is a second set of vertical mirror planes, at $45°$ to the first set (σ and σ'). They are not of the same class as σ and σ', from (7.4); we shall identify them as a second class of symmetry planes, designated individually as σ_d and σ_d'. Hence the complete set of symmetry operations for C_{4v}, placed in classes, is

$$I$$
$$C_4, C_4^3 \ (=C_4^{-1})$$
$$C_2 \ (=C_4^2)$$
$$\sigma_v, \sigma_v'$$
$$\sigma_d, \sigma_d'$$

or, more concisely, as

$$I \quad 2C_4 \quad C_2 \quad 2\sigma_v \quad 2\sigma_d$$

It is not necessary always to work out similarity transformations in order to assign symmetry operations to classes. Operations belong to the same class if they are equivalent in the normally accepted sense, as with σ_v and σ_v' under four-fold rotational symmetry. Thus, if we consider the carbonate ion, CO_3^{2-}, as another example, point group D_{3h}, we find the operations

$$I, C_3, C_3^2, C_2, C_2', C_2'', S_3, S_3^2, \sigma_h, \sigma_v, \sigma_v', \sigma_v''$$

which can be arranged as

$$I \quad 2C_3 \quad 3C_2 \quad 2S_3 \quad \sigma_h \quad 3\sigma_v$$

7.3.6 Representations

Let us begin by considering the water molecule. Its point group is C_{2v}, and we can set up reference axes according to the conventions given in Section 7.3.4. The operations of the point group C_{2v} are given by the following group multiplication table. We may note that σ_v is the plane xz, normal to y, and σ_v' is the plane yz,

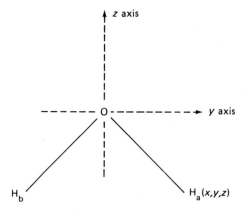

normal to x.

	I	C_2	σ_v	σ'_v
I	I	C_2	σ_v	σ'_v
C_2	C_2	I	σ'_v	σ_v
σ_v	σ_v	σ'_v	I	C_2
σ'_v	σ'_v	σ_v	C_2	I

It is an Abelian group of order 4. We can show how the symmetry operators of the group act on a point x, y, z, which could be the spatial coordinates of one of the hydrogen atoms, H_a:

$$I \quad \text{No change}$$
$$C_2 \quad x \rightarrow -x$$
$$y \rightarrow -y$$
$$z \rightarrow \text{no change}$$
$$\sigma_v \quad x, z \rightarrow \text{no change}$$
$$y \rightarrow -y$$
$$\sigma'_v \quad x \rightarrow -x$$
$$y, z \rightarrow \text{no change}$$

These results may be restated thus:

$$I.x = +1.x \quad I.y = +1.y \quad I.z = +1.z$$
$$C_2.x = -1.x \quad C_2.y = -1.y \quad C_2.z = +1.z$$
$$\sigma_v.x = +1.x \quad \sigma_v.y = -1.y \quad \sigma_v.z = +1.z$$
$$\sigma'_v.x = -1.x \quad \sigma'_v.y = +1.y \quad \sigma'_v.z = +1.z$$

or, tabulated,

C_{2v}	I	C_2	σ_v	σ_v'
x	1	-1	1	-1
y	1	-1	-1	1
z	1	1	1	1

The effect of the symmetry operations on the product functions x^2, y^2, z^2, xy, xz and yz can be obtained in a like manner. Thus, x^2, y^2 and z^2 transform like z under all operations of the group, xz transforms like x and yz like y; xy is different. Thus:

$$I \cdot xy = +1 \cdot xy$$

$$C_2 \cdot xy = +1 \cdot xy$$

$$\sigma_v \cdot xy = -1 \cdot xy$$

$$\sigma_v' \cdot xy = -1 \cdot xy$$

In all, there are four ways in which functions may transform under C_{2v}. Tabulating them, we have

C_{2v}	I	C_2	σ_v	σ_v'	
A_1	1	1	1	1	z, x^2, y^2, z^2
A_2	1	1	-1	-1	xy
B_1	1	-1	1	-1	x, xz
B_2	1	-1	-1	1	y, yz

There are no other ways in which other functions can transform; for example, x^3 transforms like x, and x^4 like x^2. The four transformations are called the *irreducible representations* of C_{2v}, the table is the *character table* of the group and the numbers in the table are the *characters of the irreducible representation* of C_{2v}. The A and B symbols are conventional labels, but with a further meaning to be considered later. Thus, we can say that x belongs to the B_1 representation of C_{2v}; x is *symmetric* with respect to I and σ_v, and *antisymmetric* with respect to C_2 and σ_v'; z is *totally symmetric* under all operations of C_{2v}.

If we form another function, say $x^2 + y^3 + z^2$, we can see that, in its entirety, it does not belong to any of the irreducible representations of C_{2v}. However, x^2 transforms as A_1, y^3 as y (B_1) and z^2 as A_1. So the function is made up of the irreducible representations $2A_1 + B_2$.

We can show that each of the representations in the character table is a true representation of the group by setting out a multiplication table for each representation. Taking A_2, for example, we have:

A_2	1	1	-1	-1
1	1	1	-1	-1
1	1	1	-1	-1
-1	-1	-1	1	1
-1	-1	-1	1	1

If we now compare this table with the group multiplication table for C_{2v}, we see that wherever σ_v or σ'_v appears in the group multiplication table, -1 appears in the same position in the A_2 table. Similar results can be obtained for the other representations. We can confirm that xy belongs to the irreducible representation A_2 because characters can be multiplied just like operators. Thus, x belongs to B_1 and y belongs to B_2; so xy belongs to $B_1 . B_2$:

	I	C_2	σ_v	σ'_v	
B_1	1	-1	1	-1	
B_2	1	-1	-1	1	
$B_1 . B_2$	1	1	-1	-1	$= A_2$

It follows that the p_x, p_y and p_z orbitals (Fig. 7.2) transform under C_{2v} as B_1, B_2

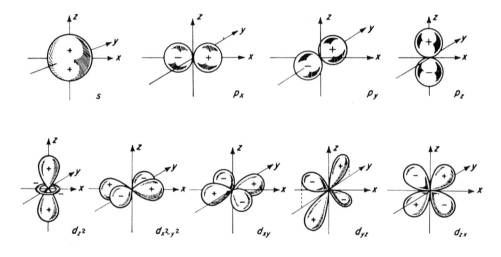

Fig. 7.2—Diagrammatic representation of the angular parts of s, p and d wave functions (atomic orbitals).

and A_1 respectively. The d-orbitals (Fig. 7.2) are a little more complex. Consider d_{xz} (d_{zx} in the figure): to which of the four representations does it belong?

$$I \;\; \rightarrow \text{no change}$$
$$C_2 \rightarrow -1$$
$$\sigma_v \;\; \rightarrow \;\;\; 1$$
$$\sigma'_v \;\; \rightarrow -1$$

so that d_{xz} belongs to the B_1 representation. In a simlar manner, we can determine the results for all the d-orbitals, leading to the following fuller character table for C_{2v}:

C_{2v}	I	C_2	σ_v	σ_v'		
A_1	1	1	1	1	z	$x^2 - y^2, z^2$
A_2	1	1	-1	-1		xy
B_1	1	-1	1	-1	x	xz
B_2	1	-1	-1	1	y	yz

We may now ask if the set of characters 3 1 1 3 can be a representation of C_{2v}. It should be clear that it is not an irreducible representation, but the numbers form the characters of a *reducible representation*, as follows:

A_1	1	1	1	1
A_1	1	1	1	1
B_2	1	-1	-1	1
$2A_1 + B_2$	3	1	1	3

In this last example, we were able to perform the reduction by a fairly simple inspection. In more complex examples, a *reduction formula* can be applied. If N is the number of times that a given irreducible representation appears in a reducible representation, then N is given by

$$N = \frac{1}{p} \Sigma X_r . X . S \qquad (7.6)$$

where p is the order of the group, X_r and X_i are, respectively, the characters of the reducible and irreducible representations, S is the number of symmetry operations in the symmetry class, and the sum is taken over all symmetry classes. Applying (7.6) to the last example, we have:

$$N_{A_1} = \tfrac{1}{4}[(3 \times 1 \times 1) + (1 \times 1 \times 1) + (1 \times 1 \times 1) + (3 \times 1 \times 1)]$$
$$= 2 \qquad (7.7)$$

$$N_{A_2} = \tfrac{1}{4}[(3 \times 1 \times 1) + (1 \times 1 \times 1) + (1 \times -1 \times 1) + (3 \times -1 \times 1)]$$
$$= 0 \qquad (7.8)$$

Similarly, we find $N_{B_1} = 0$ and $N_{B_2} = 1$, leading again to $2A_1 + B_2$ for the irreducible representation. We may write this result:

C_{2v}	I	C_2	σ_v	σ_v'	
Γ	3	1	1	3	$= 2A_1 + B_2$

where Γ is a label for the reducible representation.

So far, we have considered only those representations that can be termed non-degenerate.

7.3.7. Degenerate representations

Consider the pentafluoroantimonate ion (Fig.3.31(b)). Its point group is C_{4v}, and we can collect the symmetry operations of this group into the following classes:

$$I \quad 2C_4 \quad C_2 \,(=C_4^2) \quad 2\sigma_v \quad 2\sigma_d$$

We take the z reference axis along the apical Sb–F bond. The operation of any of the symmetry operations of C_{4v} on this bond is symmetric, so that z belongs to the totally symmetric, or A_1 representation of ths group:

C_{4v}	I	$2C_4$	C_2	$2\sigma_v$	$2\sigma_d$	
A_1	1	1	1	1	1	z

The x and y axes are directed along two of the Sb–F bonds set at right angles to each other and to the z axis. The operation of C_4 on x transforms it to y. So the p_x and p_y orbitals on antimony are interconverted by C_4 (also by σ_d). It follows that the p_x and p_y orbitals of antimony belong to a degenerate (E) representation of C_{4v}; they have the same energy. This result may be compared with that for C_{2v}, where p_x belongs to B_1 and p_y belongs to B_2.

The interconversion of x and y under the operations C_4 and σ_d in this group is a feature best represented through the use of matrices, using just the two dimensions x and y. Thus:

$$\begin{vmatrix} 0 & \bar{1} \\ 1 & 0 \end{vmatrix} \times \begin{vmatrix} x \\ y \end{vmatrix} = \begin{vmatrix} -y \\ x \end{vmatrix} \tag{7.9}$$

$$C_4$$

a result that we have encountered before (Section 5.3.3). The operation C_4^3 is the same as C_4^{-1}, or just C_4^-

$$\begin{vmatrix} 0 & 1 \\ \bar{1} & 0 \end{vmatrix} \times \begin{vmatrix} x \\ y \end{vmatrix} = \begin{vmatrix} y \\ \bar{x} \end{vmatrix} \tag{7.10}$$

$$C_4^-$$

whereas that for C_4^2 is the same as that for C_2:

$$\begin{vmatrix} \bar{1} & 0 \\ 0 & \bar{1} \end{vmatrix} \times \begin{vmatrix} x \\ y \end{vmatrix} = \begin{vmatrix} -x \\ -y \end{vmatrix} \tag{7.11}$$

$$C_2$$

The *character*, or *trace*, of a matrix is the sum of the terms on the principal diagonal. Thus, the character of C_4 and C_4^- is 0, and that of C_2 is -2. This result illustrates further why $C_4^2\,(=C_2)$ is of a symmetry class different from C_4 and C_4^3. Thus, we find that the character table tests the character of the matrices that represent all of the symmetry operations of the corresponding group.

In the example of a non-degenerate representation, the matrices were all single numbers, and each such number was, in fact, the trace of the corresponding 1×1 matrix. We can continue the construction of all the matrices for the operations of C_{4v}, and determine the characters:

C_{4v}	I	C_4	C_4^3	C_2	σ_v	σ_v'	σ_d	σ_d'
	2	0	0	-2	0	0	0	0

or, collected together,

C_{4v}	I	$2C_4$	C_2	$2\sigma_v$	$2\sigma_d$
E	2	0	-2	0	0

We may note here that operations of the same character are not necessarily in the same symmetry class, but operations in one and the same symmetry class must have the same character. The symbol E is used to represent two-fold degeneracy and, in general, the degeneracy of a representation is given by the character of its identity matrix; the degenerate functions are enclosed by parentheses in the character table. A character table for C_{4v} has the following form:

C_{4v}	I	$2C_4$	C_2	$2\sigma_v$	$2\sigma_d$		
A_1	1	1	1	1	1	z	$x^2 + y^2, z^2$
A_2	1	1	1	-1	-1		
B_1	1	-1	1	1	-1		$x^2 - y^2$
B_2	1	-1	1	-1	1		xy
E	2	0	-2	0	0	(x, y)	(xz, yz)

The tetrahedral group T_d is of great importance, since it is the point group for any regular tetrahedral arrangement, such as the C–H bonds in methane or the Cu–O bonds in the $[Cu(H_2O)_4]^{2+}$ ion.

The symmetry operations of any regular tetrahedral arrangement can be readily appreciated by inscribing it in a cube (the tetrahedron belongs to the cubic system), as in Fig. 7.3. From this figure, we can deduce the classes of symmetry operations,

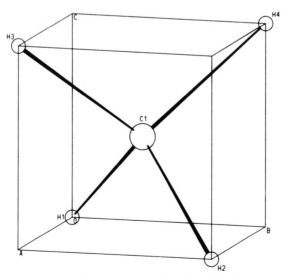

Fig. 7.3—Regular tetrahedral molecule (CH_4) inscribed in a cube: the C_4 axes of the cube are degraded to S_4 in methane, the C_3 and C_2 axes of the cube are present, along $\langle 111 \rangle$ and $\langle 110 \rangle$ respectively, the diagonal symmetry planes, normal to $\langle 110 \rangle$, are present, but both the symmetry planes normal to $\langle 100 \rangle$ and the centre of symmetry of the cube are absent in methane.

as follows:

$$I \quad 8C_3 \quad 3C_2 \quad 6S_4 \quad 6\sigma_d$$

and the character table is

T_d	I	$8C_3$	$3C_2$	$6S_4$	$6\sigma_d$		
A_1	1	1	1	1	1		$x^2 + y^2 + z^2$
A_2	1	1	1	-1	-1		
E	2	-1	2	0	0		$(2z^2 - x^2 - y^2, x^2 - y^2)$
T_1	3	0	-1	1	-1		
T_2	3	0	-1	-1	1	(x, y, z)	(xy, xz, yz)

The order of the group is 24, and the T representations are three-fold degenerate.

PROBLEMS 7

1. Draw stereograms to show the point groups S_4, S_6, D_{3h} and D_{2d}.
2. Which pairs of the operators C_2, C_3, σ and I commute?
3. Draw up a group multiplication table for

   (a) C_{2h}
   (b) C_{3v}

   What is the order of each group? Is either of them Abelian?
4. Arrange the symmetry operators of the point group in classes and give the order of the group for (a) the PF_5 molecule, (b) the $[PtCl_4]^{2-}$ ion.
5. Draw up a character table for point group C_{3v}. Is 4 1 -2 an irreducible representation in C_{3v}? If not, reduce it, and express it in terms of the representations in the character table.
6. Reduce 3 3 1 1 in C_{2v}, and 4 1 0 0 2 in T_d to irreducible representations in their groups.
7. Express the multiplication table for C_{2h} (Problem 3) and D_{2d} in the Hermann–Mauguin notation.
8. By means of similarity transformations, establish the symmetry classes for point group D_6.
9. Determine the similarity transform of S_6^5 by σ_v (σ_v is $\perp y$).
10. To what representations would we assign the function $x + y + z^2$ in (a) C_{2v}, (b) D_{3h}, (c) C_{4v}.

8

Some Applications of Group Theory in Chemistry

... but his cup of joy is full
when the results of studies
immediately find practical application

René Jules Dubos, 1986: *Louis Pasteur, Free Lance of Science*

8.1 INTRODUCTION

We shall complete our short discussion of group theory by considering some chemical applications of the principles brought out in the previous chapter, extending the arguments where necessary.

Hydrogen-like, one-electron atomic orbitals (wave functions) are used to construct wave functions for chemical species, and to describe the geometry of molecules and ions. It is important to show theoretically the equivalence of four C–H bonds in methane, or that of three B–F bonds in boron trifluoride, because without this equivalence the observed symmetries of these molecules could not be explained.

In the components of methane, we have just the $1s$ electron on each hydrogen atom, and for carbon we have $1s^2$, $2s^2$ and $2p^2$ electrons present. Those in the $1s^2$ orbital remain as a relatively 'insert' core on the carbon atom, and bonding takes place through the $2s^2$ and $2p^2$ electrons. The four tetrahedral bonds of methane can be used as a basis set of functions for describing a representation of T_d the point group of methane. By reducing this representation and comparing the result with the irreducible representations of point group T_d, we can see how a particular choice of atomic orbitals can lead to the specified symmetry of methane.

In another case, boron trifluoride, we shall find that more than one combination of atomic orbitals would lead to the observed trigonal symmetry of this molecule. In such a case, other chemical evidence must be brought to bear on the problem, so as to choose the most suitable of the results.

8.2 TRANSFORMATION PROPERTIES OF ATOMIC ORBITALS

We can use atomic orbitals as basis functions for a chemical species, determine their transformation properties, and so form irreducible representations under the point group of the given species.

In the case of $1s$ orbitals, the wave function has the form e^{-r}, where r is the radial coordinate. This function is neither even nor odd, but it may be written as

$$e^{-r} = \frac{(e^{-r} + e^r)}{2} + \frac{(e^{-r} - e^r)}{2} \tag{8.1}$$

which is the sum of an even function and an odd function respectively. Thus, for spherical symmetry (s-orbitals), *even* and *odd* are the two irreducible representations.

A sphere represents an infinite group: it contains an infinite number of rotation axes and mirror planes, the operation of any of which leave it unchanged. Atomic orbitals may be classified according to their behaviour under symmetry operations, C_R and C_i for example. Rotation is important in angular momentum problems, and gives rise to the quantum numbers l and m_l, but this topic is not discussed herein. The operation C_i relates every point x, y, z to an equivalent point at $-x$, $-y$, $-z$.

The p-orbitals are usually expressed in functional forms, such as $r\, e^{-r/2} \sin \theta\, e^{i\phi}$, where θ and ϕ are angular coordinates. The p wave functions can be expressed in an equivalent manner in rectangular coordinates, so that we find

$$2p_x = x\, e^{-r/2}$$

$$2p_x = y\, e^{-r/2}$$

$$2p_z = z\, e^{-r/2}$$

The angular representations of the s, p and d atomic orbitals are shown in Fig. 7.2. Since radial symmetry is totally symmetric, the functions p_x, p_y and p_z transforms as x, y and z respectively; (See also Section 7.3.6.) similarly the d wave functions transform as their subscripts. We shall consider two examples.

8.2.1 Boron trifluoride

Boron trifluoride belongs to point group D_{3h}; this group is of order 12. A triplet of vectors, l_1, l_2 and l_3, directed along the three B–F bonds, can be used as a basis to

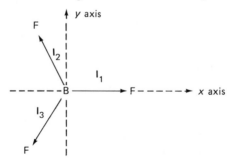

generate a reducible representation of the point group. The operations of D_{3h} (as with the CO_3^{2-} ion, Section 7.3.5) are:

$$I \quad C_3^+ \quad C_3^- \quad C_2 \quad C_2' \quad C_2'' \quad \sigma_h \quad S_3^+ \quad S_3^- \quad \sigma_v \quad \sigma_v' \quad \sigma_v''$$

(C_3^+ and C_3^- are other ways of writing C_3 and C_3^{-1}, respectively, so that $C_3^{+2} = C_3^-$). Collected into classes, they become

$$I \quad 2C_3 \quad 3C_2 \quad \sigma_h \quad 2S_3 \quad 3\sigma_v$$

We now need the character of the matrix for each symmetry class; this may be equated to the number of l vectors that are invariant under the symmetry operation.

Thus, for I we have the matrix

$$\begin{bmatrix} 1 & 0 & 0 \\ 0 & 1 & 0 \\ 0 & 0 & 1 \end{bmatrix} \cdot \begin{bmatrix} l_1 \\ l_2 \\ l_3 \end{bmatrix} = \begin{bmatrix} l_1 \\ l_2 \\ l_3 \end{bmatrix} \tag{8.2}$$

and its character is 3. For C_3, we have

$$\begin{bmatrix} 0 & 0 & 1 \\ 1 & 0 & 0 \\ 0 & 1 & 0 \end{bmatrix} \cdot \begin{bmatrix} l_1 \\ l_2 \\ l_3 \end{bmatrix} = \begin{bmatrix} l_3 \\ l_1 \\ l_2 \end{bmatrix} \tag{8.3}$$

and its character is 0. Any one of the vertical mirror planes that pass through boron and one of the fluorine atoms leaves one l vector invariant, so the character of any σ_v must be one. Thus, we can build up the following representation:

D_{3h}	I	$2C_3$	$3C_2$	σ_h	$2S_3$	$3\sigma_v$
Γ	3	0	1	3	0	1

Working along the lines discussed in the previous chapter, we can construct a character table for D_{3h}, with the result

D_{3h}	I	$2C_3$	$3C_2$	σ_h	$2S_3$	$3\sigma_v$		
A_1'	1	1	1	1	1	1		$x^2+y^2,\ z^2$
A_2'	1	1	-1	1	1	-1		
E'	2	-1	0	2	-1	0	(x, y)	$(x-y^2, xy)$
A_1''	1	1	1	-1	-1	-1		
A_2''	1	1	-1	-1	-1	1	z	
E''	2	-1	0	-2	1	0		(xz, yz)

where the primes are used when there is present more than one species of a given representation. It is clear that Γ is not an irreducible representation of D_{3h}, and we must reduce it according to the rules already established. Using (7.6), we have:

$$N_{A_1'} = \tfrac{1}{12}[3 + 0 + 3 + 3 + 0 + 3] = 1$$

Similarly, we find $N_{A_2'} = 0$, $N_{E'} = 1$, and so on, such that we obtain $\Gamma = A_1' + E'$. From the character table, we see that the orbitals available are for A_1' either s or d_z^2, and for E' we have either p_x and p_y, or $d_{x^2-y^2}$ and d_{xy}.

Group theory will not distinguish between these two sets of trigonal orbitals for boron trifluoride; from a point of view of symmetry they are equally acceptable. However, from a chemical standpoint, since boron has no d orbitals available for bonding, we select, quite correctly, the s, p_x and p_y orbitals to form sp^2 hybrids for the sigma, in-plane B–F bonds; we shall not discuss pi-bonding.

8.2.2 Methane

Methane, CH_4, belongs to the cubic point group T_d; we have referred to this species and given the character table in Section 7.3.7. We can use the four C–H bond vectors (Fig. 7.3) as a basis to generate a representation in this point group. We need next the characters of the matrices for each of the symmetry classes $I, 8C_3, 3C_2, 6S_4$ and $6\sigma_d$. Proceedings as before, we obtain

$$\Gamma = 4 \quad 1 \quad 0 \quad 0 \quad 2$$

From the character table, this representation is clearly reducible, and application of (7.6) shows that $\Gamma = A_1 + T_2$, and corresponds to either sp^3 or sd^3. Chemically, we select sp^3 as the more appropriate for carbon.

8.3 LCAO APPROXIMATIONS

The solution of the wave equation for chemical species involving more than one electron is carried out by approximation methods. The most usual method is to construct molecular orbitals by the linear combination of atomic orbitals (LCAO). Each molecular orbital is compounded as ψ_i, given by

$$\psi_1 = \sum_j c_{ij}\phi_j \tag{8.4}$$

where the coefficients c_{ij} multiplying each atomic wave function ϕ_i are adjusted so as to minimize the total electronic energy of the species. The treatment of this problem, by variational methods, is well documented in books on theoretical chemistry, and is not discussed here.

Let us return to the water molecule of symmetry C_{2v}, discussed at the beginning of Section 7.3.6. Fig. 8.1 is a conventional energy correlation diagram for the water molecule. Reference to the character table for C_{2v} shows that we can label each atomic orbital according to its symmetry. Thus, for the oxygen atom, we have

$$s: A_1 \quad p_x: B_1 \quad p_y: B_2 \quad p_z: A_1$$

For the hydrogen atoms, we must, of course, use both $1s$ atomic orbitals. Using the two O–H bond vectors as a basis to generate a representation, we find

C_{2v}	I	C_2	σ_v	σ_v'
Γ	2	0	2	0

which reduces to $\Gamma = A_1 + B_1$. We can now label the correlation diagram with the appropriate irreducible representations in the columns for the free atoms. The linear combinations for the hydrogen atoms can be written as

$$\psi_+ = \frac{1}{\sqrt{2}}[\phi_a(1s) + \phi_b(1s)] \tag{8.5}$$

and

$$\psi_- = \frac{1}{\sqrt{2}}[\phi_a(1s) - \phi_b(1s)] \tag{8.6}$$

where $1/\sqrt{2}$ is a normalization constant. We may refer to ψ_+ and ψ_- as group orbitals, or symmetry orbitals; ψ_+ is symmetric to all operations of C_{2v}, and thus is an A_1 representation, whereas ψ_- is symmetric to I and σ_v, antisymmetric to C_2 and σ_v' and is a B_1 representation.

Now, oxygen $2p_x$ and hydrogen ψ_- combine to form a bonding molecular

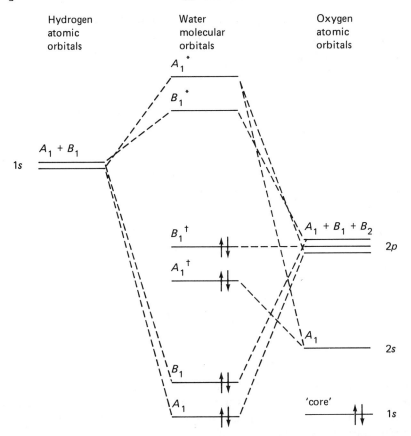

Fig. 8.1—Energy correlation diagram for the water molecule: we can consider ψ_- lying above ψ_+ for hydrogen, and oxygen $2p_x$, $2p_y$ and $2p_z$ lying above one another, in that order (this is diagrammatic; the p-orbitals are degenerate). The arrows ↑↓ represent a pair of electrons with opposed spins. There is a certain contribution to the bonding from the $2s$ orbitals because the H-O-H angle is greater than $90°$ (*ca* $104.5°$)

orbital of B_1 symmetry; it is shown diagrammatically in the central column of Fig. 8.1. The corresponding antibonding molecular orbital is B_1^*, obtained by the subtraction $2p_x - \psi_-$. In a similar manner, we can form A_1 and A_1^* from oxygen $2p_z$ and ψ_+. Atomic orbitals from different atoms will form linear combinations only if they are of the same irreducible representation. It follows that the $2p_y$ orbital of oxygen will not interact with the atomic orbitals of hydrogen; it is non-bonding, labelled $B_1^\dagger$. The other non-bonding molecular orbital, $A_1^\dagger$, is obtained from the $2s$ orbital of oxygen. Since molecular orbitals are built up, like atomic orbitals, starting with the lowest energies, we find a pair of electrons, of opposed spins, in A_1, B_1, $A_1^\dagger$ and $B_1^\dagger$. The $1s^2$ electrons of oxygen remain as a relatively inert core, significantly unaffected by the bonding process.

As a second and final example in this section, we look at a typical octahedral complex ion, such as $[Fe(CN)_6)]^{4-}$. The molecular orbital concept of metal-ligand bonding enables the complex to be discussed qualitatively in terms of orbital overlap. The atomic orbitals in a metal from the first transition series may be subscribed from the $3d$, $4s$ and $4p$ energy levels of the atom.

The symmetry of the hexacyanoferrate(II) ion is O_h $(m3m)$, the highest cubic symmetry, and the character table is as shown below:

O_h	I	$8C_3$	$6C_2$	$6C_4$	$3C_2' (=3C_4^2)$	C_i	$6S_4$	$8S_6$	$3\sigma_h$	$6\sigma_d$	
A_{1g}	1	1	1	1	1	1	1	1	1	1	$x^2+y^2+z^2$
A_{2g}	1	1	-1	-1	1	1	-1	1	1	-1	
E_g	2	-2	0	0	2	2	0	-1	2	0	$(2z^2-x^2-y^2, x^2-y^2)$
T_{1g}	3	0	-1	1	-1	3	1	0	-1	-1	
T_{2g}	3	0	1	-1	-1	3	-1	0	-1	1	(xy, xz, yz)
A_{1u}	1	1	1	1	1	-1	-1	-1	-1	-1	
A_{2u}	1	1	-1	-1	1	-1	1	-1	-1	1	
E_u	2	-1	0	0	2	-2	0	1	-2	0	
T_{1u}	3	0	-1	1	-1	-3	-1	0	1	1	(x, y, z)
T_{2u}	3	0	1	-1	-1	-3	1	0	1	-1	

From this table, we find:

$$3d \qquad x^2-y^2, z^2 \qquad E_g$$
$$xy, xz, yz \qquad T_{2g}$$
$$4s \qquad x^2+y^2+z^2 \qquad A_{1g}$$
$$4p \qquad x, y, z \qquad T_{1u}$$

The subscript notation g and u (German: *gerade* = even, *ungerade* = odd) is employed where a centre of symmetry (C_i) is present in the group, to indicate the behaviour of a representation with respect to this symmetry operation. The simplest group containing a centre of symmetry is C_i:

$C_i(=S_2)$	I	C_i
I	I	C_i
C_i	C_i	I

$C_i(=S_2)$	I	C_i	
A_g	1	1	$(x^2, y^2, z^2, xy, xz, yz)$
A_u	1	-1	(x, y, z)

The six Fe–CN bond vectors can be used as a basis for a representation of the sigma bonds in O_h. The representation is lengthy to work out, and the result only is given here:

O_h	I	$8C_3$	$6C_2$	$6C_4$	$3C_2'$	C_i	$6S_4$	$8S_6$	$3\sigma_h$	$6\sigma_d$
Γ	6	0	0	2	2	0	0	0	4	2

Following previous methods, we can reduce this representation to

$$\Gamma = A_{1g} + E_g + T_{1u}$$

We can now set up an energy correlation diagram (Fig. 8.2). If the cyanide ion ligand symmetry orbitals be denoted $\phi_1, \phi_2, \phi_3, \phi_4, \phi_5$ and ϕ_6 with respect to the directions $+x, -x, +y, -y, +z$ and $-z$, then the hybrid orbitals, with their representations, are as tabulated on the following page. Note that, with the d_{z^2} orbital, the probability along the z axis is twice that in either the x or the y axial directions.

Iron(II) Fe^{2+}
ion atomic
orbitals

Hexacyanoferrate(II)
[Fe(CN)$_6$]$^{4-}$ ion
molecular orbitals

Cyanide ion
CN$^-$ atomic
orbitals

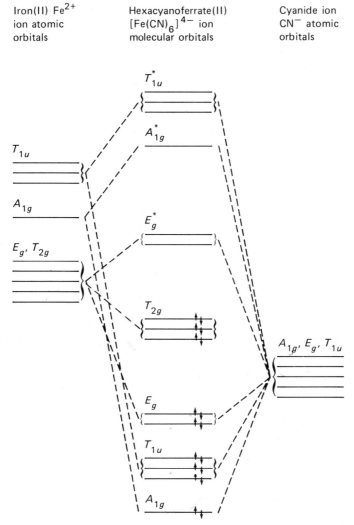

Fig. 8.2—Energy correlation diagram for the octahedral complex ion, [Fe(CN)$_6$]$^{4-}$: bonding electron pairs, shown by ↑↓, occupy the lowest six molecular orbitals of the ion, with the non-bonding electron pairs in the T_{2g} molecular orbital.

Metal atom orbitals	Normalized ligand symmetry orbitals	Representation
s	$1/\sqrt{6}(\phi_1 + \phi_2 + \phi_3 + \phi_4 + \phi_5 + \phi_6)$	A_{1g}
p_x	$1/\sqrt{2}(\phi_1 - \phi_2)$	T_{1u}
p_y	$1/\sqrt{2}(\phi_3 - \phi_4)$	T_{1u}
p_z	$1/\sqrt{2}(\phi_5 - \phi_6)$	T_{1u}
d_{z^2}	$1/\sqrt{12}(2\phi_5 + 2\phi_6 - \phi_1 - \phi_2 - \phi_3 - \phi_4)$	E_g
$d_{x^2-y^2}$	$1/2(\phi_1 + \phi_2 - \phi_3 - \phi_4)$	E_g

Three d-orbitals on the metal atom, d_{xy}, d_{xz} and d_{yz}, of representation T_{2g}, have no matching representations among the orbitals of the cyanide ion; thus, these orbitals are non-bonding

There are eighteen electrons in the molecular orbitals of the complex ion, so each of the molecular orbitals A_{1g}, T_{1u}, E_g and T_{2g} contains a pair of electrons with opposed spins. Where two symmetry orbitals of very different energy are combined, the molecular orbitals obtained are different in energy. The lower energy bonding molecular orbitals assume the character of the ligand, and we think of the twelve ligand electrons being 'donated' to the metal atom. The six electrons from the metal are then placed in the T_{2g} non-bonding orbitals. This description is a little simplistic, and a more precise treatment demands a quantitative approach.

The energy gap between the T_{2g} and the E_g^* levels is of considerable interest in the ligand-field theory of coordination compounds. In a free metal ion, the d-orbitals have a total degeneracy of five. If the metal ion is placed at the centre of octahedrally disposed negative centres (ligands) lying on the x, y and z axes, the d-orbitals are split into a group of two (E_g) and a group of three (T_{2g}) of lower energy. The d_{z^2} orbital can be resolved into a linear combination of $d_{z^2-x^2}$ and $d_{z^2-y^2}$ ($2z^2 - x^2 - y^2$), each of which is spatially equivalent to $d_{x^2-y^2}$, but such that d_{z^2} has a probability of twice that of $d_{x^2-y^2}$. The d_{xy}, d_{xz} and d_{yz} orbitals are also equivalent. The energy separation between the E_g and the T_{2g} levels is referred to as Δ, or $10Dq$, in ligand-field theory. The E_g levels lie at $\frac{3}{5}\Delta$ above the unsplit level and the T_{2g} at $\frac{2}{5}\Delta$ below it, so that a symmetry exists between the product of the fraction of Δ and the number of levels for the two states. The splitting is of significance in relation to the magnetic, spectroscopic and thermodynamic properties of coordination compounds, and the reader is referred to standard works on this subject for further study.

8.4 SELECTION RULES FOR ATOMS AND MOLECULES

Atoms have spherical symmetry, and atomic wave functions may be either odd or even according to their behaviour under C_i symmetry. The wave functions that we have used, which are of general interest, are s, p_x, p_y, p_z, d_{z^2}, $d_{x^2-y^2}$, d_{xy}, d_{xz} and d_{yz}. The s functions are spherically symmetric, and the angular parts of the p and d functions are illustrated in Fig. 7. 2. It is clear that symmetry C_i operates on these functions as follows:

$$s \qquad \text{even, } g$$
$$p \qquad \text{odd, } u$$
$$d \qquad \text{even, } g$$

For the ground state and each excited state of a chemical species there exists a potential energy curve which possesses a stable equilibrium position for the atoms of the species. Consider an electron transition between two energy levels, represented by ψ_1 and ψ_2, of energy separation $\Delta\varepsilon$, where $\Delta\varepsilon = h\nu$. It can be shown that the transition $\psi_1 \to \psi_2$ will occur only if one of the integrals

$$\int \psi_1 a \psi_2 \, d\tau \neq 0 \tag{8.7}$$

where a is x, y or z. The integral of an odd function is zero,

$$\int_{-l}^{l} z^3 \, dz = 0 \tag{8.8}$$

whereas the integral of an even function is finite:

$$\int_{-l}^{l} z^2 \, dz = 2l^3/3 \tag{8.9}$$

Since x, y and z are odd functions, the product $\psi_1\psi_2$ must be odd if $\psi_1 a\psi_2$ is to be even, but even if $\psi_1 a\psi_2$ is to be odd. Hence, if ψ_1 is even, ψ_2 must be odd, and the selection rule is

$$\text{even } (g) \leftrightarrow \text{odd } (u) \tag{8.10}$$

This rule holds for atoms, and for molecules with a centre of symmetry. For non-centrosymmetric molecules, selection rules may be derived in the following manner.

Consider again the water molecule, symmetry C_{2v}. Let a state ψ_1 belong to the irreducible representation A_1, and an excited state ψ_2 belong to B_1; the character table (Section 7.3.6) shows that $A_1 \cdot B_1 = B_1$. We need to consider integrals of the type

$$\int \psi_1 a\psi_2 \, d\tau \qquad (a = x, y \text{ or } z) \tag{8.11}$$

For $a = x$, which belongs to B_1, the integral will be non-zero and an electronic transition is permitted. If ψ_2 belongs to B_2, $A_1 \cdot B_2 = B_2$, and a transition is forbidden. We see here that, in deriving these selection rules, we do not have to evaluate the integrals; we need only know the symmetry properties of the irreducible representations.

PROBLEMS 8

1. Given that the one-electron p wave functions in radial coordinates are

$$p_1 = N_1 r e^{-r/2} \sin \theta \, e^{i\phi}$$
$$p_2 = N_2 r e^{-r/2} \sin \theta \, e^{-i\phi}$$
$$p_3 = N_3 r e^{-r/2} \cos \theta$$

   obtain the functions in terms of rectangular coordinates, given

$$x = r \sin \theta \cos \phi$$
$$y = r \sin \theta \sin \phi$$
$$z = r \cos \theta$$

2. Ammonia belongs to point group C_{3v}. Generate a representation for the group, and reduce it as necessary.

3. Continue Problem 2, so as to derive the most likely combination of atomic orbitals for the ammonia molecule; the H–N–H bond angle is ca. 107°.

4. From the results in Section 8.2.2, set up an energy correlation diagram for the methane molecule.

5. Determine the irreducible representations of the probable hybrid orbitals for the sigma (in-plane) bonds of the $[NiF_4]^{2-}$ ion; its point group is D_{4h}, with the following character table:

D_{4h}	I	$2C_4$	C_2	$2C'_2$	$2C''_2$	C_i	$2S_4$	σ_h	$2\sigma_v$	$2\sigma_d$		
A_{1g}	1	1	1	1	1	1	1	1	1	1		$x^2 + y^2, z^2$
A_{2g}	1	1	1	-1	-1	1	1	1	-1	-1		
B_{1g}	1	-1	1	1	-1	1	-1	1	1	-1		$x^2 - y^2$
B_{2g}	1	-1	1	-1	1	1	-1	1	-1	1		xy
E_g	2	0	-2	0	0	2	0	-2	0	0		(xz, yz)
A_{1u}	1	1	1	1	1	-1	-1	-1	-1	-1		
A_{2u}	1	1	1	-1	-1	-1	-1	-1	1	1	z	
B_{1u}	1	-1	1	1	-1	-1	1	-1	-1	1		
B_{2u}	1	-1	1	-1	1	-1	1	-1	1	-1		
E_u	2	0	-2	0	0	-2	0	2	0	0	(x, y)	

6. Determine whether or no the transitions between T_2 and A_1^* and between A_1 and A_1^* in methane are permitted or forbidden.

7. Determine the transformation properties for the p-orbitals of nitrogen in ammonia (C_{3v}). How does the result affect the degeneracy of the p-orbitals?

Appendix 1

Stereoviewing

The representation of crytals and molecules by stereoscopic pairs of drawings is an important aid to their study and understanding. Computer programs are available that can be used to prepare precise stereoscopic diagrams from structural data. The two diagrams of a given object must correspond to the views seen, one by each eye, in normal vision. The correct viewing of stereoscopic illustrations requires that each eye sees only the appropriate drawing; there are several ways in which stereopsis can be achieved.

(1) A stereoscope can be purchased for a modest sum; two suppliers are:
  (a) C. F. Casella & Company Limited, Regent House, Britannia Walk, London N1 7NO. This maker supplies two grades of stereoscopes.
  (b) Taylor-Merchant Corporation, 25 West 45th Street, New York, NY 10036, USA.

(2) The unaided eyes can be trained to defocus, so that each eye looks at only the appropriate diagram; the eyes must be relaxed and directed straight ahead. A thin white card held edgeways between the two views aids this direct viewing process. The stereoscopic image appears centrally between the two diagrams.

(3) A stereoscope can be constructed with a little effort. Two biconvex or planoconvex lenses, each of the same focal length (in the region of 100 mm), and diameter 20 mm to 30 mm, are mounted in a frame such that the centres of the lenses are separated by the interocular distance of about 60 mm. The frame must be shaped so that it can be held close to the eyes. Two pieces of thin, black card, prepared as shown in Fig. A1.1, can be used to support the lenses.

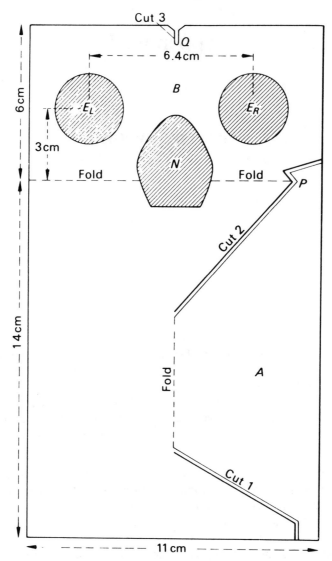

Fig. A1.1—Construction of a simple stereoscope. Cut out two pieces of card as shown, and discard the shaded portions. Make cuts along the double lines. Glue the two cards together with the two lenses E_L and E_R in positon, fold the portions A and B backwards and fix P into the cut at Q. View from the side marked B. (A similar stereoviewer is marketed by the Taylor-Merchant Corporation, New York.)

Appendix 2

Symmetry Notations‡

It should not be too much to hope that a single symmetry notation of general applicability can be devised. Without engaging in the difficult argument of whether or no the Hermann–Mauguin notation is superior to that of Schoenflies, the following tabulations and notes are offered for consideration by those who have to teach symmetry, for it may be that the impetus for change must come from this sector.

The changes proposed are not great, but they lead to a more straightforward comparison between the Hermann–Mauguin and Schoenflies notations, and are a first step towards a common notation. The implication of the suggested scheme is that it should be a modified Hermann–Mauguin notation, because of the value of this notation in both point groups *and* space groups. In the case of two-dimensional groups, the symbol *r* (reflexion line) is used in place of *m*, and the rotational degrees *R* are written in italics, so as to distinguish clearly between two-dimensional and three-dimensional point groups.

Three-dimensional crystallographic point groups

Suggested notation	Hermann–Mauguin notation	Schoenflies notation
1	1	C_1
$\bar{1}$	$\bar{1}$	C_i, S_2
2†	2	C_2
m†	*m*	C_s, S_1
2/*m*†	2/*m*	C_{2h}
222	222	D_2

‡ See also M. F. C. Ladd, Computer-assisted methods in the study of point-group symmetry, *International Journal of Mathematical Education in Science and Technology*, Vol. 7, pp. 395–400, 1976.

$2mm$‡	$mm2$	C_{2v}
mmm	mmm	D_{2h}
3	3	C_3
$\bar{3}$	$\bar{3}$	S_6
32	32	D_3
$3m$	$3m$	C_{3v}
$\bar{3}m$	$\bar{3}m$	D_{3d}
4	4	C_4
$\bar{4}$	$\bar{4}$	S_4
$4/m$	$4/m$	C_{4h}
422	422	D_4
$4mm$	$4mm$	C_{4v}
$\bar{4}2m$	$\bar{4}2m$	D_{2d}
$\frac{4}{m}mm$	$\frac{4}{m}mm$	D_{4h}
6	6	C_6
$\bar{6}$	$\bar{6}$	C_{3h}
$6/m$	$6/m$	C_{6h}
622	622	D_6
$6mm$	$6mm$	C_{6v}
$\bar{6}2m$	$\bar{6}m2$	D_{3h}
$\frac{6}{m}mm$	$\frac{6}{m}mm$	D_{6h}
23	23	T
$m3$	$m3$	T_h
$m3$	$m3$	T_h
432	432	O
$\bar{4}3m$	$\bar{4}3m$	T_d
$m3m$	$m3m$	O_h

Two-dimensional point groups

Suggested notation	Hermann–Mauguin notation
1	1
2	2
r	*m*
2rr	*2mm*
3	3
3r	*3m*
4	4
4rr	*4mm*
6	6
6rr	*6mm*

† 2 is along the *z* axis, and *m* is the *xy* plane.

‡ The first position in the symbol refers to the *z* axis (and symmetry equivalent directions), like other symbols, assuming that trigonal crystals are referred to hexagonal axes (which is a common practice).

Appendix 3

The Stereographic Projection of a Circle is a Circle

Consider Fig. A3.1. Let AB be the trace of a small circle, centre O, on a vertical section of a sphere of arbitrary radius. Lines such as PA or AX generate a cone about the axis PQ. The trace AX of its right section is one diameter of an ellipse; AB is the

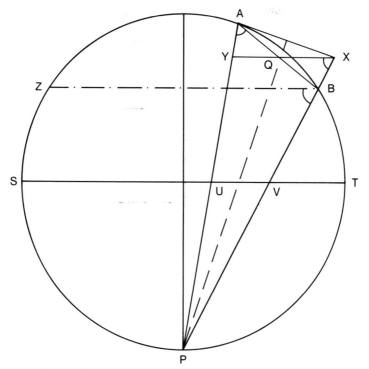

Fig. A3.1—Construction to show that the stereographic projection of a circle is a circle.

trace of one circular section of this ellipse. Symmetrically inclined to the axis PQ is the conjugate circular section, trace XY. Draw BZ parallel to the primitive ST. Then,

$$\widehat{ZBP} = \widehat{PAB} \quad \text{(equal arcs } PZ, PB) \tag{A3.1}$$

But triangles PAB and PXY are congruent, so that

$$\widehat{PAB} = \widehat{PXY} \, (= \widehat{ZBP}) \tag{A3.2}$$

Hence

$$XY \parallel BZ \, (\parallel ST) \tag{A3.3}$$

and, therefore, UV is a diameter of a circle on the primitive. It follows that all angular relationships on a crystal are reproduced on the stereogram, so that it becomes a symmetry-true (angle-true) representation.

Appendix 4

Analytical Geometry of Direction Cosines

A4.1 DIRECTION COSINES OF A LINE

In Fig. A4.1, let P_1 be any point x_1, y_1, z_1 referred to rectangular x, y and z axes. Draw lines from P_1 perpendicular to the x, y and z axes to cut them at A, B and C respectively. Thus, $OA = x_1$, $OB = y_1$ and $OC = z_1$. The direction cosines of OP are given by $\cos \chi_1 = x_1/OP_1$, $\cos \psi_1 = y_1/OP_1$ and $\cos \omega_1 = z_1/OP_1$. Hence,

$$\cos^2 \chi_1 + \cos^2 \psi_1 + \cos^2 \omega_1 = (x_1^2 + y_1^2 + z_1^2)/OP_1^2 \tag{A4.1}$$

But since x_1, y_1 and z_1 are the projections of OP_1 on to the x, y and z axes, it follows that

$$x_1^2 + y_1^2 + z_1^2 = OP_1^2 \tag{A4.2}$$

Hence,

$$\cos^2 \chi_1 + \cos^2 \psi_1 + \cos^2 \omega_1 = 1 \tag{A4.3}$$

A4.2 ANGLE BETWEEN TWO LINES

On another line OP_2, let a point x_2, y_2, z_2 be marked off such that the lengths of OP_1 and OP_2 are equal, say r. We have for OP_1, from above,

$$x_1 = r \cos \chi_1, \quad y_1 = r \cos \psi_1 \quad \text{and} \quad z_1 = r \cos \omega_1$$

and for OP_2,

$$x_2 = r \cos \chi_2, \quad y_2 = r \cos \psi_2 \quad \text{and} \quad z_2 = r \cos \omega_2$$

If the origin is shifted from O to P_1, then the coordinates of P_2 become

$$x_2' = x_2 - x_1, \quad y_2' = y_2 - y_1, \quad z_2' = z_2 - z_1 \tag{A4.4}$$

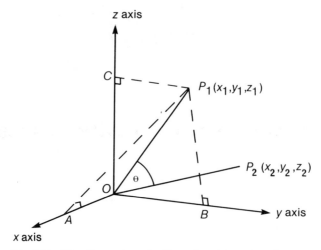

Fig. A4.1—Direction cosines of a line, referred to rectangular axes.

so that the length P_1P_2 is given by

$$(P_1P_2)^2 = (x_2 - x_1)^2 + (y_2 - y_1)^2 + (z_2 - z_1)^2$$
$$= r^2(\cos^2 \chi_1 + \cos^2 \chi_2 - 2 \cos \chi_1 \cos \chi_2$$
$$+ \cos^2 \psi_1 + \cos^2 \psi_2 - 2 \cos \psi_1 \cos \psi_2$$
$$+ \cos^2 \omega_1 + \cos^2 \omega_2 - 2 \cos \omega_1 \cos \omega_2) \qquad \text{(A4.5)}$$

Using (A4.3),

$$(P_1P_2)^2 = 2r^2[1 - (\cos \chi_1 \cos \chi_2 + \cos \psi_1 \cos \psi_2 + \cos \omega_1 \cos \omega_2)] \qquad \text{(A4.6)}$$

In the isosceles triangles OP_1P_2

$$P_1P_2/2 = r\sin(\theta/2) \qquad \text{(A4.7)}$$

Therefore,

$$(P_1P_2)^2/(2r^2) = \sin^2(\theta/2) = 1 - \cos \theta \qquad \text{(A4.8)}$$

Comparing (A4.6) and (A4.8), we obtain

$$\cos \theta = \cos \chi_1 \cos \chi_2 + \cos \psi_1 \cos \psi_2 + \cos \omega_1 \cos \omega_2 \qquad \text{(A4.9)}$$

Appendix 5

Transformation and Rotation Matrices

The matrix for a three-fold rotation in the xy plane, normal to the z axis, used in Solution 3.14, may be obtained by the following general method.

Consider the point $P(x, y, z)$ with respect to the hexagonal axes (Fig. A5.1). The rectangular coordinates are given by

$$x_0 = \frac{\sqrt{3}}{2}x$$

$$y_0 = -\tfrac{1}{2}x + y \qquad\qquad (A5.1)$$

$$z_0 = z$$

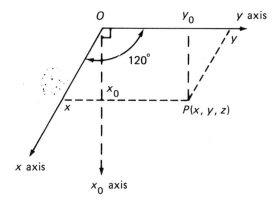

Fig. A5.1—Transformation of coordinates: hexagonal to rectangular axes.

Next we need to rotate the point x_0, y_0, z_0 through and angle ϕ about the z axis. Consider Fig. A5.2: the coordinates of P_0 may be written as

$$x_0 = r \cos \theta$$

$$y_0 = r \sin \theta \qquad \text{(A5.2)}$$

$$z_0$$

where r is the distance of P_0 from the origin. After an anticlockwise rotation of P_0 through an angle ϕ to P'_0, we have

$$x'_0 = r \cos(\theta + \phi)$$

$$y'_0 = r \sin(\theta + \phi) \qquad \text{(A5.3)}$$

$$z'_0$$

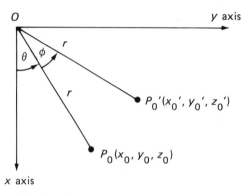

Fig. A5.2—Rotational transformation in two dimensions.

Expanding the trigonometric functions and using (A5.2) we obtain

$$x'_0 = x_0 \cos \phi - y_0 \sin \phi$$

$$y'_0 = x_0 \sin \phi + y_0 \cos \phi \qquad \text{(A5.4)}$$

$$z'_0$$

For $\phi = 120°$, (A5.4) becomes

$$x'_0 = -\tfrac{1}{2}x_0 - \tfrac{\sqrt{3}}{2}y_0$$

$$y'_0 = \tfrac{\sqrt{3}}{2}x_0 - \tfrac{1}{2}y_0 \qquad \text{(A5.5)}$$

$$z'_0$$

After this rotation, the resulting coordinates are transformed back to hexagonal axes, using the inverse of (A5.1), namely

$$x' = \frac{2}{\sqrt{3}}x'_0$$

$$y' = \tfrac{1}{2}x' + y_0 = \frac{1}{\sqrt{3}}x'_0 + y'_0 \qquad \text{(A5.6)}$$

$$z' = z_0$$

We could carry out the transformations of (A5.1) to (A5.6) so as to obtain the desired result. However, there is a neater method, using matrices, as follows.

Equations (A5.1), (A5.5) and (A5.6) may be represented by matrices **T**, **R** and $\mathbf{T}^{-1}$ respectively, where $\mathbf{T}^{-1}$ is the inverse of **T**, such that

$$\mathbf{T} = \begin{vmatrix} \frac{\sqrt{3}}{2} & 0 & 0 \\ -\frac{1}{2} & 1 & 0 \\ 0 & 0 & 1 \end{vmatrix} \tag{A5.7}$$

$$\mathbf{R} = \begin{vmatrix} -\frac{1}{2} & -\frac{\sqrt{3}}{2} & 0 \\ \frac{\sqrt{3}}{2} & -\frac{1}{2} & 0 \\ 0 & 0 & 1 \end{vmatrix} \tag{A5.8}$$

$$\mathbf{T}^{-1} = \begin{vmatrix} \frac{2}{\sqrt{3}} & 0 & 0 \\ \frac{1}{\sqrt{3}} & 1 & 0 \\ 0 & 0 & 1 \end{vmatrix} \tag{A5.9}$$

Following the sequence of operations, and Section 3.9, we have

$$\mathbf{T}^{-1} \cdot \mathbf{R} \cdot \mathbf{T} \cdot \mathbf{x} = \mathbf{x}' \tag{A5.10}$$

$$(\mathbf{R} \cdot \mathbf{T}) = \begin{vmatrix} 0 & -\frac{\sqrt{3}}{2} & 0 \\ 1 & -\frac{1}{2} & 0 \\ 0 & 0 & 1 \end{vmatrix} \tag{A5.11}$$

and

$$\mathbf{T}^{-1} \cdot (\mathbf{R} \cdot \mathbf{T}) = \begin{vmatrix} 0 & -1 & 0 \\ 1 & -1 & 0 \\ 0 & 0 & 1 \end{vmatrix} \tag{A5.12}$$

Thus, (A5.12) gives $\bar{y}$, $x - y$, z for a three-fold rotation about the z axis, from the point x, y, z, both referred to hexagonal axes.

The mirror reflexion of the point $\bar{y}$, $x - y$, z across the plane $\perp y$ may be considered in a like manner. Thus, x_0, y_0, z_0 (rectangular axes) reflects to $x_0, \bar{y}_0, z_0$, and the appropriate matrix is

$$M = \begin{array}{|ccc|} 1 & 0 & 0 \\ 0 & \bar{1} & 0 \\ 0 & 0 & 1 \end{array} \qquad (A5.13)$$

So we need

$$\mathbf{T}^{-1} \cdot \mathbf{M} \cdot \mathbf{R} \cdot \mathbf{T} \cdot \mathbf{x} = \mathbf{x}'' \qquad (A5.14)$$

or

$$\mathbf{T}^{-1} \cdot \mathbf{M} \cdot \mathbf{T} \cdot \mathbf{x}' = \mathbf{x}'' \qquad (A5.15)$$

The reader may satisfy himself that both (A5.14) and (A5.15), carried out according to the rules already described, lead to $\bar{y}, \bar{x}, z$ for $\mathbf{x}''$. It now follows that the m plane relating x, y, z to $\bar{y}, \bar{x}, z$ is a vertical plane bisecting the angle between the x and $\bar{y}$ axes, that is, the m plane normal to the third (u) axis, and of the same *form* as the m plane given initially.

Matrix Multiplication

For convenience, we state here the rule for multiplication of two matrices. If $\mathbf{A}$ and $\mathbf{B}$ are given by

$$\mathbf{A} = \begin{array}{|ccc|} a_{11} & a_{12} & a_{13} \\ a_{21} & a_{22} & a_{23} \\ a_{31} & a_{32} & a_{33} \end{array} \qquad \mathbf{B} = \begin{array}{|ccc|} b_{11} & b_{12} & b_{13} \\ b_{21} & b_{22} & b_{23} \\ b_{31} & b_{32} & b_{33} \end{array} \qquad (A5.16)$$

and their product is $\mathbf{C}$, where

$$\mathbf{A} \cdot \mathbf{B} = \mathbf{C} \qquad (A5.17)$$

then

$$c_{11} = a_{11}b_{11} + a_{12}b_{21} + a_{13}b_{31} \qquad (A5.18)$$

or, generally,

$$c_{ij} = \sum_{k=1}^{3} a_{ik}b_{kj} \qquad (5.19)$$

Appendix 6

Simple Vector Methods

A6.1 SUM, DIFFERENCE AND DOT (SCALAR) PRODUCT OF TWO VECTORS

Let $\mathbf{r}_1$ and $\mathbf{r}_2$ be any two vectors with respect to a common origin O (Fig. A6.1). Draw QS perpendicular to OP to meet OP in S. From the extension of Pythagoras' theorem, we have in triangle OPQ

$$PQ^2 = OQ^2 + OP^2 - 2\,OP\,OS \tag{A6.1}$$

or

$$PQ^2 = OQ^2 + OP^2 - 2\,OP\,OQ\cos\widehat{QOS} \tag{A6.2}$$

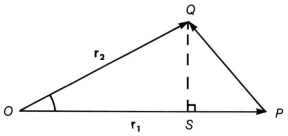

Fig. A6.1—Vectors $\mathbf{r}_1$ and $\mathbf{r}_2$ from a common origin, and the meaning of $r_1 \cdot r_2$.

Using vector quantities, (A6.2) becomes

$$PQ^2 = r_1^2 + r_2^2 - 2r_1r_2\cos\widehat{\mathbf{r}_1\mathbf{r}_2} \tag{A6.3}$$

In vector notation

$$\mathbf{PQ} = \mathbf{r}_2 - \mathbf{r}_1 \tag{A6.4}$$

and the dot product of each side with itself is

$$\mathbf{PQ} \cdot \mathbf{PQ} = (\mathbf{r}_2 - \mathbf{r}_1) \cdot (\mathbf{r}_2 - \mathbf{r}_1) \tag{A6.5}$$

Multiplication must take place along the same line, so that each pair of terms multiplied is modified by the cosine of the angle between them. Thus, we obtain

$$PQ^2 = r_1^2 + r_2^2 - 2\mathbf{r}_1 \cdot \mathbf{r}_2 \tag{A6.6}$$

Since (A6.3) and (A6.6) are identical, it follows that the dot (scalar) product of the two vectors $\mathbf{r}_1$ and $\mathbf{r}_2$ is given by

$$\mathbf{r}_1 \cdot \mathbf{r}_2 = r_1 r_2 \cos \widehat{r_1 r_2} \tag{A6.7}$$

It is a simple matter to obtain an analogous expression for the sum of the two vectors $\mathbf{r}_1$ and $\mathbf{r}_2$. These results are of general applicability, and we have used some of them in Section 4.4.2.

A6.2 BOND LENGTH AND ANGLE CALCULATIONS

Let P, Q and R (Fig. A6.2) be three atoms of fractional coordinates $x_1, y_1, z_1, x_2, y_2, z_2$ and x_3, y_3, z_3 respectively. We shall calculate the bond lengths PQ and PR and the angle $\widehat{QPR}$. We have seen from (A6.4) that the bond length PQ may be obtained from

$$\mathbf{PQ} = \mathbf{r}_2 - \mathbf{r}_1 \tag{A6.8}$$

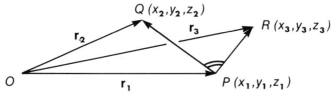

Fig. A6.2—Vectors $\mathbf{r}_1$, $\mathbf{r}_2$ and $\mathbf{r}_3$ from a common origin, and bond length and bond angle calculations.

Expressing $\mathbf{r}_1$ and $\mathbf{r}_2$ in terms of their components, we obtain

$$\mathbf{r}_1 = x_1\mathbf{a} + y_1\mathbf{b} + z_1\mathbf{c} \tag{A6.9}$$

and

$$\mathbf{r}_2 = x_2\mathbf{a} + y_2\mathbf{b} + z_2\mathbf{c} \tag{A6.10}$$

so that

$$\mathbf{r}_2 - \mathbf{r}_1 = (x_2 - x_1)\mathbf{a} + (y_2 - y_1)\mathbf{b} + (z_2 - z_1)\mathbf{c} \tag{A6.11}$$

From (A6.6) and (A6.7)

$$\begin{aligned} PQ = [(x_2 - x_1)^2 a^2 &+ (y_2 - y_1)^2 b^2 + (z_2 - z_1)^2 c^2 \\ &+ (y_2 - y_1)(z_2 - z_1)bc \cos \alpha \\ &+ (z_2 - z_1)(x_2 - x_1)ca \cos \beta \\ &+ (x_2 - x_1)(y_2 - y_1)ab \cos \gamma]^{1/2} \end{aligned} \tag{A6.12}$$

A similar expression may be deduced for the second bond length PR by replacing x_2, y_2 and z_2 in (A6.12) by x_3, y_3 and z_3 respectively.

The bond angle is, from (A6.7), given by

$$\cos \widehat{QPR} = \frac{\mathbf{PQ} \cdot \mathbf{PR}}{PQ\ QR} \tag{A6.13}$$

so that we may write

$$\cos \widehat{QPR} = [(x_2 - x_1)\mathbf{a} + (y_2 - y_1)\mathbf{b}$$
$$+ (z_2 - z_1)\mathbf{c}] \cdot [(x_3 - x_1)\mathbf{a} + (y_3 - y_1)\mathbf{b} + (z_3 - z_1)\mathbf{c}]/PQ\ QR$$
$$= [(x_2 - x_1)(x_3 - x_1)a^2 + (y_2 - y_1)(y_3 - y_1)b^2$$
$$+ (z_2 - z_1)(z_3 - z_1)c^2$$
$$+ 2(y_2 - y_1)(z_3 - z_1)bc \cos \alpha$$
$$+ 2(z_2 - z_1)(x_3 - x_1)ca \cos \beta$$
$$+ 2(x_2 - x_1)(y_3 - y_1)ab \cos \gamma]/(PQ\ QR)$$
$$\tag{A6.14}$$

where PQ is given by (A6.12) and QR, by (A6.12) with the above-mentioned replacement of $\mathbf{x}_2$ by $\mathbf{x}_3$. Equations (A6.12) and (A6.14) are quite general. Simplification arises for a higher symmetry system: for example, in the cubic system (A6.12) becomes

$$PQ = a[(x_1 - x_2)^2 + (y - y_2)^2 + (z_1 - z_2)^2]^{1/2} \tag{A6.15}$$

A6.3 CROSS PRODUCT OF TWO VECTORS

The cross product $\mathbf{a} \times \mathbf{b}$ is defined to be $\mathbf{i}\ ab \sin \theta$, where $\mathbf{i}$ is a unit vector perpendicular to $\mathbf{a}$ and $\mathbf{b}$, and in a direction such that $\mathbf{a}$, $\mathbf{b}$ and $\mathbf{i}$ form a right-handed set of directions; θ is the angle of rotation of $\mathbf{a}$ to $\mathbf{b}$. Thus, $\mathbf{a} \cdot \mathbf{b} = \mathbf{b} \cdot \mathbf{a}$ (a scalar), $\mathbf{a} \times \mathbf{b} = -\mathbf{b} \times \mathbf{a}$ (a vector).

A6.4 VOLUME OF A PARALLELEPIPEDON

The volume of a parallelepipedon of edge vectors $\mathbf{a}$, $\mathbf{b}$ and $\mathbf{c}$ is given by

$$V = \mathbf{a} \cdot (\mathbf{b} \times \mathbf{c}) \tag{A6.16}$$

Let $\mathbf{a}$, $\mathbf{b}$ and $\mathbf{c}$ be expressed in terms of an orthogonal set of unit vectors $\mathbf{i}$, $\mathbf{j}$ and $\mathbf{k}$, such that

$$\mathbf{a} = a_1\mathbf{i} + a_2\mathbf{j} + a_3\mathbf{k} \tag{A6.17}$$
$$\mathbf{b} = b_1\mathbf{i} + b_2\mathbf{j} + b_3\mathbf{k} \tag{A6.18}$$
$$\mathbf{c} = c_1\mathbf{i} + c_2\mathbf{j} + c_3\mathbf{k} \tag{A6.19}$$

Then, following Section A6.3, and equations (A6.7) and (A6.16), we obtain

$$V = \mathbf{a} \cdot (\mathbf{b} \times \mathbf{c})$$
$$= (a_1\mathbf{i} + a_2\mathbf{j} + a_3\mathbf{k}) \cdot (b_1c_2\mathbf{k} + b_1c_3\mathbf{j} + b_2c_1\mathbf{k} + b_2c_3\mathbf{i} + b_3c_1\mathbf{j} + b_3c_2\mathbf{i}) \tag{A6.20}$$

The right-hand side of (A6.20) is the expansion of the determinant

$$V = \begin{vmatrix} a_1 & a_2 & a_3 \\ b_1 & b_2 & b_3 \\ c_1 & c_2 & c_3 \end{vmatrix} \tag{A6.21}$$

Since we can interchange rows and columns of a determinant without changing its value, we can write

$$V^2 = \begin{vmatrix} a_1 & a_2 & a_3 \\ b_1 & b_2 & b_3 \\ c_1 & c_2 & c_3 \end{vmatrix} \cdot \begin{vmatrix} a_1 & b_1 & c_1 \\ a_2 & b_2 & c_2 \\ a_3 & b_3 & c_3 \end{vmatrix} \tag{A6.22}$$

Multiplying the determinants according to the rule for matrices (A5.19), we obtain

$$V^2 = \begin{vmatrix} a_1a_1 + a_2a_2 + a_3a_3 & a_1b_1 + a_2b_2 + a_3b_3 & a_1c_1 + a_2c_2 + a_3c_3 \\ b_1a_1 + b_2a_2 + b_3a_3 & b_1b_1 + b_2b_2 + b_3b_3 & b_1c_1 + b_2c_2 + b_3c_3 \\ c_1a_1 + c_2a_2 + c_3a_3 & c_1b_1 + c_2b_2 + c_3b_3 & c_1c_1 + c_2c_2 + c_3c_3 \end{vmatrix} \tag{A6.23}$$

Reference to (A6.17)–(A6.19) shows that (A6.23) is

$$V^2 = \begin{vmatrix} \mathbf{a} \cdot \mathbf{a} & \mathbf{a} \cdot \mathbf{b} & \mathbf{a} \cdot \mathbf{c} \\ \mathbf{b} \cdot \mathbf{a} & \mathbf{b} \cdot \mathbf{b} & \mathbf{b} \cdot \mathbf{c} \\ \mathbf{c} \cdot \mathbf{a} & \mathbf{c} \cdot \mathbf{b} & \mathbf{c} \cdot \mathbf{c} \end{vmatrix} \tag{A6.24}$$

Evaluating (A6.24), we obtain

$$V^2 = a^2b^2c^2 + ab \cos \gamma\, bc \cos \alpha\, ca \cos \beta + ac \cos \beta\, ba \cos \gamma\, bc \cos \alpha$$
$$- ca \cos \beta\, b^2 ac \cos \beta - cb \cos \alpha\, bc \cos \alpha\, a^2$$
$$- c^2\, bc \cos \gamma\, ab \cos \gamma \tag{A6.25}$$

Simplifying

$$V = abc(1 - \cos^2 \alpha - \cos^2 \beta - \cos^2 \gamma + 2 \cos \alpha \cos \beta \cos \gamma)^{1/2} \tag{A6.26}$$

Solutions to Problems

SOLUTIONS 1

1. (a) One *m* plane, provided that any decoration not consistent with this symmetry is ignored;
   (b) Two *m* planes intersecting at 90°; their line of intersection is a two-fold rotation axis;
   (c) One ∞ axis;
   (d) As for (b);
   (e) Three mutually perpendicular *m* planes; there are also three two-fold rotation axes perpendicular to the *m* planes;
   (f) As for (b) if it contains a depression, or as for (e) if its pairs of opposite faces are alike;
   (g) An ∞ axis with an *m* plane perpendicular to this axis;
   (h) As for (c);
   (i) One *m* plane;
   (j) No symmetry; all faces are different.

2. (a) . .   (b) .   .

       . .   .   .

3. $\frac{1}{2}\Omega$.

4. (a) Odd   (b) Even
   (c) Even   (d) Even
   (e) Even   (f) Odd

5. F, G, J, P, Q, R.

SOLUTIONS 2

1. (a) (3$\bar{1}$0)   (b) ($\bar{3}$212)   (c) (403)
   (d) (1$\bar{4}$2)   (e) (08$\bar{1}$)   (f) (1$\bar{1}$6)

2. (a) $[\bar{1}10]$ (b) $[031]$ (c) $[012]$
   (d) $[\bar{3}01]$ (e) $[231]$ (f) $[11\bar{1}]$

3. (a) $[\bar{1}10]$ (b) $[031]$ (c) $[012]$
   (d) $[031],[231]$ (e) $[231]$ (f) $[11\bar{1}]$

4. (a) $(\bar{1}10)$ (b) $(2\bar{1}2)$ (e) $(12\bar{1})$
   (d) (101) (e) (010) (f) (112)

5. The planes must satisfy the Weiss Zone Equation, in this case, $h - k + 3l = 0$. Two possible planes are (210) and (121). (Check by the cross-multiplication rule.)

6. By the cross-multiplication rule, the zone symbol is $[k_1l_2 - k_2l_1,\ l_1h_2 - l_2h_1,\ h_1k_2 - h_2k_1]$. From the Weiss Zone Equation, we have $(k_1l_2 - k_2l_1)(mh_1 + nh_2) + (l_1h_2 - l_2h_1)(mk_1 + nk_2) + (h_1k_2 - h_2k_1)(ml_1 + nl_2) = 0$. Expansion of this equation shows that it is true.

7. Use Fig. 2.8. Let d_1 be the normal to $(h_1k_1l_1)$ from the origin of the reference axes. From triangles such as OAN, the direction cosines of d_1 are $\cos \chi_1 = d_1h_1/a$, $\cos \psi_1 = d_1k_1/b$, $\cos \omega_1 = d_1l_1/c$. Hence, from (A4.3), $h_1^2/a^2 + k_1^2/b^2 + l_1^2/c^2 = 1/d_1^2$, and $\cos \chi_1 = (h_1/a)/(h_1^2/a^2 + k_1^2/b^2 + l_1^2/c^2)^{1/2}$. Similar expressions may be obtained for ψ_1 and ω_1, and for the direction cosines of the normal d_2 to the plane $(h_2k_2l_2)$. The angle between the normals now follows from (A4.9).

8. Fig. S2.1 shows Fig. P2.1 with convenience letters added. We are given that $t = 63.90°$. Following Fig. 2.27, $c/a = \tan \widehat{001\ 101}$, whence $\widehat{001\ 101} = 61.00° = (p+q)$, and $c/(2a) = \tan \widehat{001\ 102}$, whence $p = 42.05(4°)$. The angle between q and r is the angle between the zones $[010]$ and $[\bar{1}01]$, which is $90°$ for an orthogonal crystal. In the spherical triangle $(p+q)rt$, and using (2.34), $\cos 63.90° = \cos 61.00° \cos r$, whence $r = 24.84(5)°$. In the triangle, sqt, $q = 61.00° - 42.054 = 18.94(6)°$. Hence, $\cos s = \cos 18.946° \cos 24.845°$, so that s, the angle $\widehat{111\ 102}$, is $30.86°$.

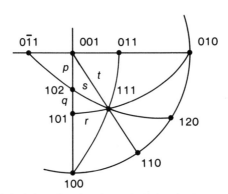

Fig. S2.1—Calculation of angles of spherical triangles (compare Fig. 2.1).

9. (a) Refer to Fig. S2.2
   (b) Refer to Fig. S2.2. The centre of the stereogram corresponds to 001, although this face does not appear on the crystal. Some zone symbols have been added; they were of use in the indexing process.
   (c) $a/b = \tan \widehat{am} = 1.000$; $c/b = \tan \widehat{001\ 011} = \tan(90° - \widehat{ae}) = 1.777$.
   (d) $\widehat{pe} = 41°04'$.

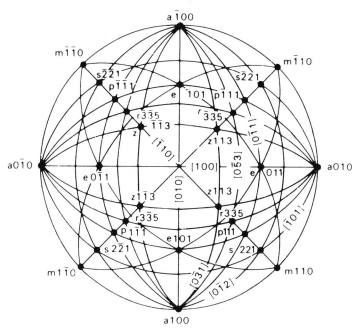

Fig. S2.2—Completed stereogram of Anatase; some zone symbols, such as [0$\bar{1}$2], have been inserted.

SOLUTIONS 3

1. (a) Refer to Figs. 3.1(c), 3.2(b) and (c).

    (b) All parts of an object must respond to the symmetry elements present. This condition can be achieved with a parallelogram only in the case of a two-fold rotation point at its centre.

2. Refer to Figs 3.17(b), 3.18(c) and 3.21(a)

    mm2: 1, 2, *m*

    $\bar{4}$2*m*: 1, 2, *m*, 222, *mm2*, $\bar{4}$

    23: 1, 2, 222, 3

    $mm2 \xrightarrow{\bar{1}} mmm$; $\bar{4}2m \xrightarrow{\bar{1}} \dfrac{4}{m}mm$; $23 \xrightarrow{\bar{1}} m3$

3. *m*: monoclinic (*y* axis unique), $m \perp y$

    422: tetragonal, 4 along *z*, 2 along *x* and *y*, 2 along [110] and [1$\bar{1}$0]

    $\bar{6}$*m*2: hexagonal, $\bar{6}$ along *z*, $m \perp x$, *y* and *u*, 2 along *x*, *y* and *u*

    $\bar{4}$3*m*: cubic, $\bar{4}$ along *x*, *y* and *z*, 3 along $\langle 111 \rangle$, $m \perp \langle 110 \rangle$

4. 522: $\widehat{5\,2} = 90°$, $\widehat{2\,2} = 36°$

    5 along *z*, 2 along *x*, *y*, *u*, *v* and *w* at 72° to each other and normal to *z*, 2 at 36° to *x*, *y*, *u*, *v* and *w* and normal to *z*

    $\bar{8}$2*m*: $\bar{8}$ along *z*, 2 along *x*, *y*, *u*, *v* at 45° to each other and normal to *z*, $\bar{2}$ (*m*) at $22\frac{1}{2}°$ to *x*, *y*, *u* and *v* and normal to *z*.

5. Refer to the stereograms in Fig. S3.1: both correspond to $\dfrac{4}{m}mm$ (see Fig. 3.18(d)).

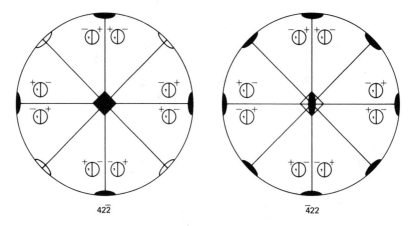

$42\bar{2}$　　　　　　　　　　　　　$\bar{4}22$

Fig. S3.1—Stereograms of the general form for point groups $42\bar{2}$ and $\bar{4}22$; they are identical— standard symbol $\dfrac{4}{m}mm$.

6. *mmm*; $\dfrac{2}{m}$.

7.

	{010}	{120}	{123}
mm2	2 (s)	4 (g)	4 (g)
4mm	4 (a)	8 (g)	8 (g)
432	6 (s)	12 (s)	24 (g)

8. (a) $1, \bar{1}$　　　　　　(b) $1, \dfrac{2}{m}$　　　　　　(c) m, mmm

　　(d) $m, \dfrac{4}{m}$　　　　(e) $2mm, \dfrac{4}{m}mm$　　(f) $m, \dfrac{4}{m}mm$

　　(g) $1, \bar{3}$　　　　　　(h) $m, \bar{3}m$　　　　　　(i) $3, \dfrac{6}{m}$

　　(j) $3m1, \dfrac{6}{m}mm$　　(k) $3, m3$　　　　　　(l) $2mm, m3m$

9. (a) 3　　(b) 2　　(c) $\bar{4}3m$　　(d) 32 or 6

10. (a) $\dfrac{6}{m}mm$　　(b) *mm2*　　(c) *mm2*

　　(d) *mm2*　　(e) *mmm*　　(f) *mm2*
　　(g) *m*　　　(h) *m*　　　(i) $\bar{6}m2$

　　(j) *mm2*　　(k) *mm2*　　(l) $\dfrac{2}{m}$

　　(m) *mm2*　　(n) $\dfrac{6}{m}mm$

11. (a) $\bar{4}3m$　　(b) *3m*　　(c) *mm2*　　(d) *m*　　(e) 1

12. (a) 2 or m
    (b) 2

13. Refer to Fig. S3.2(a) and (b). Multiplicity of general form is 20; five fluorine atoms are in special positions of symmetry $mm2$, and two fluorine atoms are in special positions of symmetry $5m$, in the point group $\bar{10}m2$.

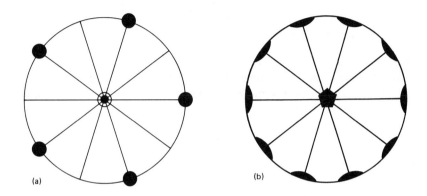

Fig. S3.2—(a) Stereogram of the bond directions in IF_7. (There is no difference in the significance of the points at the centre.) (b) Stereogram of the symmetry elements in IF_7.

14. $m \perp x$ $\bar{4}$ along z

$$
\begin{vmatrix} \bar{1} & 0 & 0 \\ 0 & 1 & 0 \\ 0 & 0 & 1 \end{vmatrix} \cdot \begin{vmatrix} 0 & 1 & 0 \\ \bar{1} & 0 & 0 \\ 0 & 0 & \bar{1} \end{vmatrix} = \begin{vmatrix} 0 & \bar{1} & 0 \\ \bar{1} & 0 & 0 \\ 0 & 0 & \bar{1} \end{vmatrix}
$$

$\qquad R_2 \qquad\qquad R_1 \qquad\qquad R_3$

Hence, $x, y, z \xrightarrow{R_2 \cdot R_1} \bar{y}, \bar{x}, \bar{z}$, or $x' = R_2 \cdot R_1 \cdot x$. Matrix R_3 refers to a diad axis at 45° to the x and y axes, and in the xy plane.

15. $m \perp y$ 3 along z

$$
\begin{vmatrix} 1 & 0 & 0 \\ 1 & \bar{1} & 0 \\ 0 & 0 & 1 \end{vmatrix} \cdot \begin{vmatrix} 0 & \bar{1} & 0 \\ 1 & \bar{1} & 0 \\ 0 & 0 & 1 \end{vmatrix} = \begin{vmatrix} 0 & \bar{1} & 0 \\ \bar{1} & 0 & 0 \\ 0 & 0 & 1 \end{vmatrix}
$$

$\qquad R_2 \qquad\qquad R_1 \qquad\qquad R_3$

Matrix R_3 corresponds to an m plane normal to the u axis, but it is of the *same form* as R_2. Thus, the m generated by the operation of 3 followed by m is coincident with the m (form) given initially, and there is no non-trivial meaning for the third position in any trigonal point-group symbol.

SOLUTIONS 4

1.   (i)   (ii)
     (a) 4mm   6mm
     (b) Square   Hexagonal
     (c) Another square   The symmetry at each point
         may be drawn as   is degraded to 2mm. A rectangular
         a p unit cell.   net is produced, and a p unit cell may be selected.
     Transformation equations in both cases are:

$$\mathbf{a}' = \mathbf{a}/2 - \mathbf{b}/2; \quad \mathbf{b}' = \mathbf{a}/2 + \mathbf{b}/2$$

2. Monoclinic; $\dfrac{2}{m}$; 2.58 nm.

3. Refer to Section 4.1.
   (i) True orthorhombic lattice; centring one pair of opposite faces is consistent
       with orthorhombic symmetry. The standard description is C, and the (unit
       cell) transformation matrix for $B \to C$ is

$$\begin{vmatrix} 0 & 0 & 1 \\ 1 & 0 & 0 \\ 0 & 1 & 0 \end{vmatrix}$$

   (ii) True lattice, but symmetry is actually orthorhombic; it is a special case with
        $a = b$ (Note that this case shows the significance of writing $a \not\subset b$. It is not a
        requirement of orthorhombic symmetry that $a = b$, but it can be an
        experimental result, within the limits of measurement.)
   (iii) Triclinic I represents a true lattice, but it may be reduced to P by the matrix
         $(I \to P)$

$$\begin{vmatrix} -\frac{1}{2} & \frac{1}{2} & \frac{1}{2} \\ \frac{1}{2} & -\frac{1}{2} & \frac{1}{2} \\ \frac{1}{2} & \frac{1}{2} & -\frac{1}{2} \end{vmatrix}$$

4.   (i) $\mathbf{a}' = -\mathbf{a} + \mathbf{c}; \ \mathbf{b}' = -\mathbf{b}; \ \mathbf{c}' = \mathbf{a}$ (remember right-handed axes and β obtuse)

     (ii) $\mathbf{a}' = \mathbf{b}/2 + \mathbf{c}/2; \quad \mathbf{b}' = \mathbf{a}/2 + \mathbf{c}/2; \quad \mathbf{c}' = \mathbf{a}/2 + \mathbf{b}/2$

     (iii) $\mathbf{a}' = \mathbf{a}/2 - \mathbf{b}/2; \quad \mathbf{b}' = \mathbf{a}/2 + \mathbf{b}/2; \quad \mathbf{c}' = \mathbf{c}$

5. The symmetry at each lattice point is $\bar{1}$; it is a special case of the triclinic system
   with $\gamma = 90°$.

6. Refer to Fig. S4.1; $\mathbf{a}$, $\mathbf{b}$ and $\mathbf{c}$ refer to the rhombohedron $V_R/V_C = 1/4$.

7. $\mathbf{a}' = -\mathbf{a}/2 + \mathbf{b}/2 + \mathbf{c}/2$

   $\mathbf{b}' = \mathbf{a}/2 - \mathbf{b}/2 + \mathbf{c}/2$

   $\mathbf{c}' = \mathbf{a}/2 + \mathbf{b}/2 - \mathbf{c}/2$

   It is straightforward to show that $a' = b' = c' = 4.58$ Å, and $\alpha' = \beta' = \gamma' = 89.1°$,
   using the methods discussed in Section 4.4.2 and Appendix 6. Since $V_R/V_I =$
   $|\mathbf{S}| = 1/2$, the volume of the rhombohedral unit cell is one-half of that of the I
   unit cell, and so it is primitive (R).

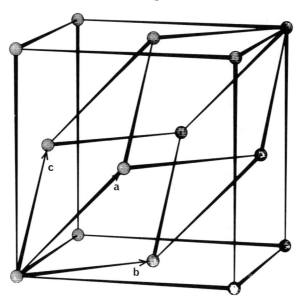

Fig. S4.1—Face-centred cubic unit cell with rhombohedron outlined within it.

8. $|\mathbf{S}|$ is 2; hence, $V_{II} = 2V_I$

$$\mathbf{S}^T = \begin{vmatrix} 1 & 1 & 0 \\ \bar{1} & 1 & 0 \\ 0 & 0 & 1 \end{vmatrix} \qquad \mathbf{S}^{-1} = \begin{vmatrix} \frac{1}{2} & \frac{1}{2} & 0 \\ -\frac{1}{2} & \frac{1}{2} & 0 \\ 0 & 0 & 1 \end{vmatrix} \qquad (\mathbf{S}^{-1})^T = \begin{vmatrix} \frac{1}{2} & -\frac{1}{2} & 0 \\ \frac{1}{2} & \frac{1}{2} & 0 \\ 0 & 0 & 1 \end{vmatrix}$$

New coordinates are, from (4.82), 0.397, -0.274, 0.314.

9. $(100):(110):(1\bar{3}0):(042) = 1:0.71:0.32:0.45$
   The most likely boundary forms are $\{100\}$ and $\{110\}$.
10. (a) Refer to Fig. S4.2.

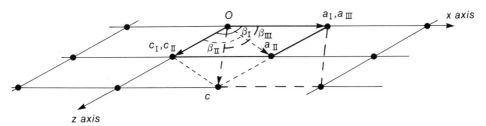

Fig. S4.2—Monoclinic cells I, II and III, drawn in projection along b.

(b) Cell I C, cell II I, cell III F
Cell III would not be chosen because it is twice the volume of standard monoclinic cell I, and II. Cell I corresponds to the conventional choice C. (Choice of cell II might be justified because it is of the same size as I and has a less obtuse β angle, which is sometimes convenient in drawings of the structure.)

(c) $\mathbf{a_{II}} = \mathbf{A} \cdot \mathbf{a_I}$ $\mathbf{A} = \begin{vmatrix} 1 & 0 & 1 \\ 0 & 1 & 0 \\ 0 & 0 & 1 \end{vmatrix}$

$\mathbf{a_{III}} = \mathbf{B} \cdot \mathbf{A_I}$ $\mathbf{B} = \begin{vmatrix} 1 & 0 & 0 \\ 0 & 1 & 0 \\ 1 & 0 & 2 \end{vmatrix}$

$\mathbf{A}^{-1} \cdot \mathbf{a_{II}} = \mathbf{A}^{-1} \cdot \mathbf{A} \cdot \mathbf{a_I} = \mathbf{a_I}$

$\mathbf{a_{III}} = \mathbf{B} \cdot \mathbf{A}^{-1} \cdot \mathbf{a_{II}} = \mathbf{C} \cdot \mathbf{a_I}$ $\mathbf{C} = \begin{vmatrix} 1 & 0 & \bar{1} \\ 0 & 1 & 0 \\ 1 & 0 & 1 \end{vmatrix}$

(d) Cell I (chosen): $\mathbf{a^*} = 0.201$ Å^{-1}, $\mathbf{b^*} = 0.0660$ Å^{-1},

$\mathbf{c^*} = 0.322$ Å^{-1}, $\beta^* = 28°17'$ (see Fig. S4.3)

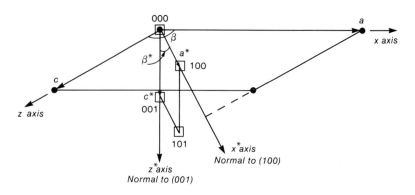

Fig. S4.3—Bravais C unit cell and corresponding reciprocal unit cell, drawn in projection along b (b^*).

(e) $\mathbf{a_{II}^*} = (\mathbf{A}^{-1})^{\mathsf{T}} \cdot \mathbf{a_I^*}$ $(\mathbf{A}^{-1})^{\mathsf{T}} = \begin{vmatrix} 1 & 0 & 0 \\ 0 & 1 & 0 \\ \bar{1} & 0 & 1 \end{vmatrix}$

Hence, $a_{II}^* = 0.201$ Å^{-1}, $b_{II}^* = 0.0660$ Å^{-1}

$c_{II}^* = 0.173$ Å^{-1} $\beta_{II}^* = 61°29'$

Calculation from cell II directly: $a_{II}^* = 0.201$ Å^{-1}, $b_{II}^* = 0.0660$ Å^{-1}, $c_{II}^* = 0.174$ Å^{-1}, $\beta^* = 61°25'$ (The discrepancy is a rounding error only.)

(f) (i) $\mathbf{a}_{\text{III}} = \mathbf{B} \cdot \mathbf{a}_{\text{I}}$; hence $(hkl)_{\text{III}} = (13\bar{3})_{\text{III}}$

(ii) $\mathbf{a}_{\text{II}} = \mathbf{C}^{-1} \cdot \mathbf{a}_{\text{III}}$; hence, $\mathbf{U}_{\text{II}} = ((\mathbf{C}^{-1})^{-1})^{\text{T}} \cdot \mathbf{U}_{\text{III}}$

$$= (\mathbf{C})^{\text{T}} \cdot \mathbf{U}_{\text{III}}$$

$$\mathbf{C} = \begin{vmatrix} 1 & 0 & 1 \\ 0 & 1 & 0 \\ \bar{1} & 0 & 1 \end{vmatrix} \quad \text{whence,} \quad [UVW]_{\text{III}} = [\bar{1}\bar{1}5]_{\text{III}}$$

(iii) $\mathbf{a}_{\text{III}} = \mathbf{B} \cdot \mathbf{a}_{\text{I}}$

$$\mathbf{x}_{\text{III}} = (\mathbf{B}^{-1})^{\text{T}} \cdot \mathbf{x}_{\text{I}} \qquad (\mathbf{B}^{-1})^{\text{T}} = \begin{vmatrix} 1 & 0 & 0 \\ 0 & 1 & 0 \\ -\frac{1}{2} & 0 & \frac{1}{2} \end{vmatrix}$$

hence, $(x, y, z)_{\text{III}} = (0.6, 0.5, -0.45)_{\text{I}}$

SOLUTIONS 5

1. See Fig. S5.1

$$(0, 0; \tfrac{1}{2}, \tfrac{1}{2}) +$$

8	(f)	1	$x, y; x, \bar{y}; \bar{x}, y; \bar{x}, \bar{y}$
4	(e)	m	$0, y; 0, \bar{y}$
4	(d)	m	$x, 0; \bar{x}, 0$
4	(c)	2	$\tfrac{1}{4}, \tfrac{1}{4}; \tfrac{1}{4}, \tfrac{3}{4}$
2	(b)	$2mm$	$0, \tfrac{1}{2}$
2	(a)	$2mm$	$0,0$

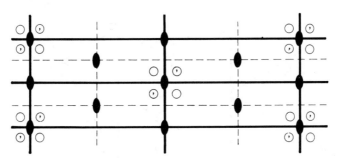

Fig. S5.1—Symmetry elements and general equivalent positions in $c2mm$.

2. See Fig. S5.2. If 2 is placed at the intersection of m and g, it does not lead to a space group: the arrangement causes a continuous halving of the repeat distance parallel to g.

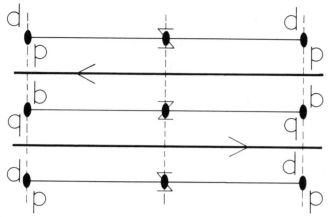

Fig. S5.2—Symmetry elements in $p2mg$; the minimum number of P, V and Z motifs have been inserted.

3. See Fig. S5.3.

$$x, y; \bar{y}, x-y; y-x, \bar{x}; \bar{y}, \bar{x}; x, x-y; y-x, y$$

4. (a)

4	e	1	$x, y, z; \bar{x}, \bar{y}, \bar{z}; x, \frac{1}{2}-y, \frac{1}{2}+z; \bar{x}, \frac{1}{2}+y, \frac{1}{2}-z$
2	(d)	$\bar{1}$	$\frac{1}{2}, 0, \frac{1}{2}; \frac{1}{2}, \frac{1}{2}, 0$
2	(c)	$\bar{1}$	$0, 0, \frac{1}{2}; 0, \frac{1}{2}, 0$
2	(b)	$\bar{1}$	$\frac{1}{2}, 0, 0; \frac{1}{2}, \frac{1}{2}, \frac{1}{2}$
2	(a)	$\bar{1}$	$0, 0, 0; 0, \frac{1}{2}, \frac{1}{2}$

(001) $p2gm$ (100) $p2gg$ (010) $p2$, $c' = c/2$

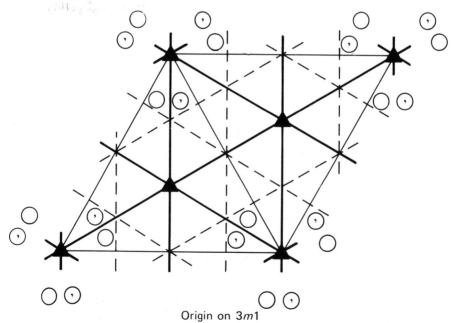

Origin on $3m1$

Fig. S5.3—Symmetry elements and general positions in $p3m1$.

(b) The molecules lie on a set of special positions, e.g. (a), with the mid-point of the central C(1)–C(1)′ bond on $\bar{1}$. It follows that the molecule is centrosymmetric and, hence, planar. The planarity indicates possible conjugation involving the C(1)–C(1)′ bond (its length is 0.149 nm).

5. See Fig. S5.4; two unit cells are shown in projection on (010). $Pa \equiv Pc$ by interchange of the x and z axes (only y is fixed by the monoclinic symmetry). $Pc \equiv Pn$ through the transformation

$$\mathbf{a}'(Pc) = \mathbf{a}(Pn); \quad \mathbf{c}'(Pc) = -\mathbf{a}(Pn) + \mathbf{c}(Pn)$$

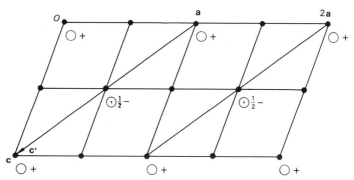

Fig. S5.4—Two monoclinic unit cells, drawn in projection along b, showing the equivalence of Pa, Pc and Pn.

6. $P2_122_1$

7. See Fig. S5.5; 2_1 along z

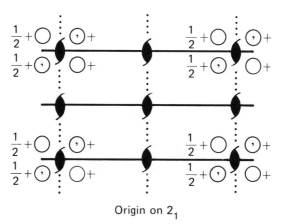

Origin on 2_1

Fig. S5.5—Symmetry elements and general positions in $Pmc2_1$ (origin on 2_1).

8. $\bar{1}$ at $0, 0, 0$
   b across (p, y, z)
   a across (x, q, z)
   n across (x, y, r)

$x, y, z \xrightarrow{\quad b \quad} 2p - x, \frac{1}{2} + y, z$

$\downarrow a$

$\frac{1}{2} + 2p - x, 2q - \frac{1}{2} - y, z$

$\downarrow n$

$2p - x, 2q - y, 2r - z$

$\xrightarrow{\quad \bar{1} \quad} \bar{x}, \bar{y}, \bar{z}$

Half-shift rule:
Total shift $= b/2 + a/2$
$+ a/2 \times b/2 \equiv 0$

Hence, $p = q = r = 0$

See Fig. S5.6 for confirmation. Full symbol $P\dfrac{2\,2\,2}{b\,a\,n}$.

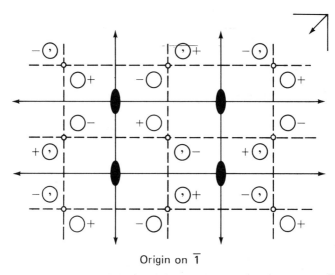

Origin on $\bar{1}$

Fig. S5.6—Symmetry elements and general positions in *Pban* (origin at $\bar{1}$).

9. See Fig. S5.7

$$x, y, z; \ x, y, \bar{z}; \ \tfrac{1}{2} - x, \tfrac{1}{2} - y, z; \ \tfrac{1}{2} - x, \tfrac{1}{2} - y, \bar{z}$$

$$\bar{x}, \tfrac{1}{2} + y, z; \ \bar{x}, \tfrac{1}{2} + y, \bar{z}; \ \tfrac{1}{2} + x, \bar{y}, z; \ \tfrac{1}{2} + x, \bar{y}, \bar{z}$$

Centre of symmetry at $\frac{1}{4}, \frac{1}{4}, 0$, for example. Transform origin to this point:

$$x - \tfrac{1}{4}, y - \tfrac{1}{4}, z; \ x - \tfrac{1}{4}, y - \tfrac{1}{4}, \bar{z}; \ \tfrac{1}{4} - x, \tfrac{1}{4} - y, z; \ \tfrac{1}{4} - x, \tfrac{1}{4} - y, \bar{z}$$

$$-x - \tfrac{1}{4}, \tfrac{1}{4} + y, z; \ -x - \tfrac{1}{4}, \tfrac{1}{4} + y, \bar{z}; \ \tfrac{1}{4} + x, -y - \tfrac{1}{4}, z; \ \tfrac{1}{4} + x, -y - \tfrac{1}{4}, \bar{z}$$

Let $x - \frac{1}{4} = x_0$, $y - \frac{1}{4} = y_0$ and $z = z_0$. Then, we have

$$x_0, y_0, z_0; \ x_0, y_0, \bar{z}_0; \ \bar{x}_0, \bar{y}_0, z_0; \ \bar{x}_0, \bar{y}_0, \bar{z}_0$$

$$\tfrac{1}{2} - x_0, \tfrac{1}{2} + y_0, z_0; \ \tfrac{1}{2} - x_0, \tfrac{1}{2} + y_0, \bar{z}_0; \ \tfrac{1}{2} + x_0, \tfrac{1}{2} - y_0, z_0; \ \tfrac{1}{2} + x_0, \tfrac{1}{2} - y_0, \bar{z}_0$$

which, dropping the subscript, are the coordinates of the general equivalent positions for *Pbam* with the origin on $\bar{1}$.

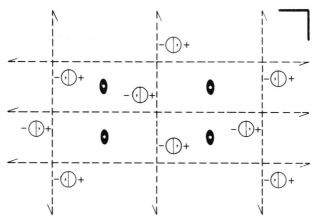

Fig. S5.7—Symmetry elements and general positions in *Pbam* (origin at *bam*).

10. The transformation is

$$\mathbf{a}' = -\mathbf{c} \qquad \mathbf{b}' = \mathbf{b} \qquad \mathbf{c}' = \mathbf{a}$$

Pban: Normal to a' is the plane that was normal to c, i.e. n glide. Normal to b' is the plane that was normal to b, i.e. a glide; but a is now c', so it becomes a c glide. Normal to c' is the plane that was normal to a, i.e. b glide. Hence, in the $\bar{\mathbf{c}}\mathbf{ba}$ setting the symbol is *Pncb*. Similarly, *Cmca* becomes *Acam*.

11. Set 4 along $[0, 0, z]$, n across (p, y, z) and c across (q, q, z)

$$x, y, z \xrightarrow{\quad n \quad} 2p - x, \tfrac{1}{2} + y, \tfrac{1}{2} + z$$
$$\Big\downarrow c$$
$$q - \tfrac{1}{2} - y, q - 2p + x, z$$
$$x, y, z \xrightarrow{\quad 4 \quad} \bar{y}, x, z$$

Thus, $q = 1/2$ and $p = q/2 = 1/4$. Hence, see Fig. S5.8.

$$x, y, z; \ \bar{y}, x, z; \ \bar{x}, \bar{y}, z; \ y, \bar{x}, z$$

$$\tfrac{1}{2} - x, \tfrac{1}{2} + y, \tfrac{1}{2} + z; \ \tfrac{1}{2} - y, \tfrac{1}{2} - x, \tfrac{1}{2} + z; \ \tfrac{1}{2} + x, \tfrac{1}{2} - y, \tfrac{1}{2} + z; \ \tfrac{1}{2} + y, \tfrac{1}{2} + x, \tfrac{1}{2} + z$$

12. See Fig. S5.9.

$$x, y, z; \ \bar{y}, x - y, z; \ y - x, \bar{x}, z; \ y, x, \tfrac{1}{2} + z; \ \bar{x}, y - x, \tfrac{1}{2} + z; \ x - y, \bar{y}, \tfrac{1}{2} + z$$

It may be noted here, and in Problem 3, that 3 followed by m (or c) introduces two more m (or c) in the same form of symmetry planes (there is no symbol 3*mm*). Trigonal symmetry illustrates a degenerate case of the symmetry combination rule (Section 3.6.3). Thus, $m_y \cdot 3 \equiv m_u$:

$$
\begin{vmatrix} 1 & 0 & 0 \\ 1 & \bar{1} & 0 \\ 0 & 0 & 1 \end{vmatrix}
\cdot
\begin{vmatrix} 0 & \bar{1} & 0 \\ 1 & \bar{1} & 0 \\ 0 & 0 & 1 \end{vmatrix}
=
\begin{vmatrix} 0 & \bar{1} & 0 \\ \bar{1} & 0 & 0 \\ 0 & 0 & 1 \end{vmatrix}
$$

$$m \perp y \qquad \bar{4} \text{ along } z \qquad m \perp u$$

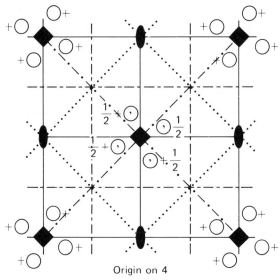

Fig. S5.8—Symmetry elements and general positions in *P4nc*.

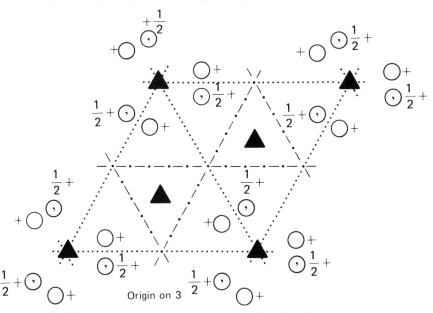

Fig. S5.9—Symmetry elements and general positions in *P31c*.

If multiplied the other way, i.e. $3 \cdot m_y$, the equivalent operator is $m \perp x$; all three planes are equivalent under the three-fold rotation, and are of one and the same form.

13. The combination $a \cdot n$ in this space group results in the matrix

$$
\begin{array}{|ccc|}
\bar{1} & 0 & 0 \\
0 & \bar{1} & 0 \\
0 & 0 & 1
\end{array}
\quad
\begin{array}{|c|}
\frac{1}{2} \\
\frac{1}{2} \\
\frac{1}{2}
\end{array}
$$

which is a 2_1 operator oriented $[\frac{1}{4}, \frac{1}{4}, z]$. In $Pna2_1$, the standard origin is with 2_1 along $[0, 0, z]$, so that the n glide would be across $(\frac{1}{4}, y, z)$, and the a glide across $(x, \frac{1}{4}, z)$. In this case, the resultant translation vectors would sum to $0, 0, \frac{1}{2}$.

14. (a) $p2$, (b) $c2mmm$. This problem illustrates the fundamental nature of the symmetry of the *contents* of the unit cell. A c cell could be chosen for the Z array, but there is no purpose in so doing, even though γ would be $90°$; the pattern still has symmetry $p2$ (although the symmetry of a plain rectangle is higher). In the H array, the c cell is the conventional choice—the *pattern* has symmetry $c2mm$. The unconventional unit cell $a = b$, $\gamma = 90°$ is also a proper unit cell, but the pattern remains of symmetry $c2mm$—it is invariant under choice of unit cell.

SOLUTIONS 6

1. From (6.1), $\lambda = 1.24/V$, where V is in kV and λ is in nm. Hence, $\lambda_{min} = 0.041$ *nm*.
2. From (6.2), $I/I_0 = 0.41$.
3. From (4.53), $1/d_{hkl}^2 = h^2/a^2 + k^2/b^2 + l^2/c^2$. Hence, for the cubic system, $d_{123} = a/\sqrt{14} = 0.168$ nm. From (6.5), $\theta_{123} = 27.4°$.
4. $4mm$; $\langle 100 \rangle$.
5. The atoms in the unit cell are related as x, y, z and $\frac{1}{2} + x, \frac{1}{2} + y, z$. Following Section 6.7.1, the simplified equation is

$$F_{hkl} = \sum_{j=1}^{N/2} f_{j,\theta} \exp[i2\pi(hx_j + ky_j + lz_j)]\left[1 + \exp\left(i2\pi \frac{(h+k+l)}{2}\right)\right]$$

Hence, the limiting conditions for an I unit cell are $hkl: h + k + l = 2n$.

6. The atoms in the unit cell are related as x, y, z and $\frac{1}{2} - x, \bar{y}, \frac{1}{2} + z$. Following Section 6.7.1 as before, and setting $h = k = 0$, we have

$$F_{00l} = \sum_{j=1}^{N/2} f_{j,\theta} \exp[i2\pi lz_j][1 + \exp(i2\pi l/2)]$$

Hence, the systematic absences are $00l: l = 2n + 1$.

7. The n glide plane relates x, y, z to $\frac{1}{2} + x, \frac{1}{2} + y, \bar{z}$, and the simplified structure factor equation becomes, with $l = 0$,

$$F_{hk0} = \sum_{j=1}^{N/2} f_{j,\theta} \exp[i2\pi(hx_j + ky_j)]\left[1 + \exp\left(i2\pi \frac{h+k}{2}\right)\right]$$

and the limiting conditions are $hk0: h + k = 2n$.

8. (a) $P2_1$, $P2_1/m$
   (b) $C2$, Cm, $C2/m$
   (c) $Pbm2$ (**ba̅c** setting of $Pma2$), $Pb2_1m$ (**bca** setting of $Pmc2_1$), $Pbmm$ (**cab** setting of $Pmma$)
   (d) $Ibca$
   (e) $I4_1/a$
   (f) $P4cc$, $P\dfrac{4}{m}cc$
   (g) We have either two molecules in general positions of $P2_1$, or two molecules in special positions in $P2_1/m$, that is, on $\bar{1}$. Since a steroid molecule has no symmetry, the second choice is precluded and the space group is $P2_1$.

(h) $hkl: h + k + l = 2n$

$0kl: k = 2n, (l = 2n)$

$h0l: l = 2n, (h = 2n)$

$hk0: h = 2n, (k = 2n)$

$h00: (h = 2n)$

$0k0: (k = 2n)$

$00l: (l = 2n)$

9. With the origin at $\bar{1}$, all atoms are related as x, y, z and $\bar{x}, \bar{y}, \bar{z}$. From (6.13) and (6.14), we obtain

$$A'_{hkl} = \sum_{j=1}^{N/2} f_{j,\theta}[\cos 2\pi(hx_j + ky_j + lz_j) + \cos 2\pi(-hx_j - ky_j - lz_j)]$$

$$= 2 \sum_{j=1}^{N/2} f_{j,\theta} \cos 2\pi(hx_j + ky_j + lz_j)$$

$$B'_{hkl} = \sum_{j=1}^{N/2} f_{j,\theta} \sin 2\pi(hx_j + ky_j + lz_j) + \sin 2\pi(-hx_j - ky_j - lz_j)]$$

$$= 0$$

From (6.12),

$$F_{hkl} = 2 \sum_{j=1}^{N/2} f_{j,\theta} \cos 2\pi(hx_j + ky_j + lz_j)$$

where the sum includes only the atoms in the asymmetric unit, that is, not related by the centre of symmetry. This equation shows the usefulness of choosing the origin at $\bar{1}$ in centrosymmetric crystals.

10. Let the geometrical structure factor be G_{hkl}. In $P2_1/c$, the general equivalent positions are $\pm[x, y, z; x, \frac{1}{2} - y, \frac{1}{2} + z]$. From the results for Problem 9, we have

$$G_{hkl} = 2\left[\cos 2\pi(hx + ky + lz) + \cos 2\pi\left(hx - ky - lz + \frac{k+l}{2}\right)\right]$$

Using the given identity,

$$G_{hkl} = 4 \cos 2\pi\left(hx + lz + \frac{k+l}{4}\right) \cos 2\pi\left(ky - \frac{k+l}{4}\right)$$

$$= 4\left[\cos 2\pi(hx + lz) \cos 2\pi\frac{(k+l)}{4} - \sin 2\pi(hx + lz) \sin 2\pi\frac{(k+l)}{4}\right]$$

$$\times \left[\cos 2\pi ky \cos 2\pi\frac{(k+l)}{4} + \sin 2\pi ky \sin 2\pi\frac{(k+l)}{4}\right]$$

$$= 4\left[\cos 2\pi(hx + lz) \cos 2\pi ky \cos^2 2\pi\frac{(k+l)}{4}\right.$$

$$\left. - \sin 2\pi(hx + lz) \sin 2\pi ky \sin^2 2\pi\frac{(k+l)}{4}\right]$$

since terms involving the mixed product $\cos 2\pi \dfrac{(k+l)}{4} \sin 2\pi \dfrac{(k+l)}{4}$ are zero (k and l are integers). We consider two cases:

$k+l=2n$: $G_{hkl} = 4\cos 2\pi(hx+lz)\cos 2\pi ky$

$k+l=2n+1$: $G_{hkl} = -4\sin 2\pi(hx+lz)\sin 2\pi ky$

Hence, the limiting conditions are:

$$h0l: l=2n; \quad 0k0: k=2n$$

We see also that $G_{h\bar{k}l} = G_{hkl}$ for $k+l$ even, but $G_{h\bar{k}l} = -G_{hkl}$ for $k+l$ odd. Since F_{hkl} follows G_{hkl} in this respect, we have that $F_{h\bar{k}l} = (-1)^{k+l}F_{hkl}$.

SOLUTIONS 7

1. Refer to Figs 3.11(i), 3.11(h), 3.20(c) and 3.18(c).
2. C_2 commutes with σ, and I commutes with all operators.

3. (a)

C_{2h}	I	C_2	C_i	σ_h	
I	I	C_2	C_i	σ_h	
C_2	C_2	I	σ_h	C_i	
C_i	C_i	σ_h	I	C_2	
σ_h	σ_h	C_i	C_2	I	Order 4; Abelian

(b)

C_{3v}	I	C_3	C_3^2	σ_v	σ_v'	σ_v''	
I	I	C_3	C_3^2	σ_v	σ_v'	σ_v''	
C_3	C_3	C_3^2	I	σ_v'	σ_v''	σ_v	
C_3^2	C_3^2	I	C_3	σ_v''	σ_v	σ_v'	
σ_v	σ_v	σ_v''	σ_v'	I	C_3^2	C_3	
σ_v'	σ_v'	σ_v	σ_v''	C_3	I	C_3^2	
σ_v''	σ_v''	σ_v'	σ_v	C_3^2	C_3	I	Order 6; non-Abelian

It is straightforward to work out these tables from the corresponding stereograms; σ_v, σ_v' and σ_v'' are taken in order anticlockwise starting from $\sigma_v \perp y$.

4. (a) $I \quad 2C_3 \quad 3C_2 \quad \sigma_h \quad 2S_3 \quad 3\sigma_v$ Order 12 (D_{3h})

(b) $I \quad 2C_4 \quad C_2 \quad 2C_2' \quad 2C_2'' \quad C_i \quad 2S_4 \quad \sigma_h \quad 2\sigma_v \quad 2\sigma_d$ Order 16 (D_{4h})

5.

C_{3v}	I	$2C_3$	$3\sigma_v$		
A_1	1	1	1	z	x^2+y^2, z^2
A_2	1	1	-1		
E	2	-1	0	(x, y)	$(x^2-y^2, xy)(xz, yz)$

4 1 -2 is reducible to $2A_2 + E$.

6. From the character table for C_{2v} in Section 7.3.6, and (7.6) if necessary, 3 3 1 1 reduces to $2A_1 + A_2$. Similarly, from the character table for T_d in Section 7.3.7, 4 1 0 0 2 reduces to $A_1 + T_2$.

7. (a)

2/m	1	2	1̄	m
1	1	2	1̄	m
2	2	1	m	1̄
m	m	1̄	2	1

(b) $\dfrac{4}{m}mm$

	1	4̄	4̄³	2	2′	2″	m	m′
1	1	4̄	4̄³	2	2′	2″	m	m′
4̄	4̄	2	1	4̄³	m	m′	2″	2′
4̄³	4̄³	1	2	4̄³	m′	m	2′	2″
2	2	4̄³	4̄	1	2″	2′	m′	m
2′	2′	m′	m	2″	1	2	4̄³	4̄
2″	2″	m	m′	2′	2	1	4̄	4̄³
m	m	2′	2″	m′	4̄	4̄³	1	2
m′	m′	2″	2′	m	4̄³	4̄	2	1

8. From Appendix 5, (A5.12), remembering that $6 = 2 \cdot 3^2$, we obtain the matrix for a C_6 rotation:

$$\begin{array}{ccc} 1 & \bar{1} & 0 \\ 1 & 0 & 0 \\ 0 & 0 & 1 \end{array}$$

C_2 is its own inverse, and the matrix for C_2 along y (hexagonal axes) is:

$$\begin{array}{ccc} 1 & 0 & 0 \\ 1 & \bar{1} & 0 \\ 0 & 0 & \bar{1} \end{array}$$

We now compute $C_2 \cdot C_6 \cdot C_2$, with the result:

$$\begin{array}{ccc} 0 & 1 & 0 \\ \bar{1} & 1 & 0 \\ 0 & 0 & 1 \end{array}$$

which is the matrix for C_6^5, or C_6^{-1}. In a similar manner, we find C_6^2 and C_6^4 (C_3 and C_3^2) are conjugate, and $C_6^3 = C_2$. Then we can work on the other symmetry operators, or by inspection of Fig. 3.20(a), leading to the symmetry clases I $2C_6$ $2C_3$ C_2 $3C_2'$ $3C_2''$.

9. The matrix for $S_6^5 (= S_6^{-1})$ will be that for the combination $C_2 \cdot C_6 \cdot C_2$ (see Solution 7.8) but with the bottom row as 0 0 1̄. That for σ_v will be that for C_2 but with the bottom row as 0 0 1. Thus, from (7.1), we find $\sigma_v \cdot S_6^5 \cdot \sigma_v = S_6$, so that S_6^5 and S_6 are conjugate.

10. From the appropriate character tables, to be found in this chapter

$C_{2v}: A_1 + B_1 + B_2$

$D_{3h}: A_1' + 2E'$

$C_{4v}: A_1 + 2E$

SOLUTIONS 8

1. p_3 becomes p_z by the substitution $z = r \cos \theta$; p_1 and p_2 are expanded ($e^{i\phi} = \cos \phi + \mathbf{i} \sin \phi$) and linear combinations formed such that p_x derives from their sum as $N_1' = x\,e^{-r/2}$, and p_y from their difference as $N_2' = y\,e^{-r/2}$.

2. Working along the lines of Section 8.2.1, the symmetry classes for C_{3v} are I, $2C_3$ and $3\sigma_v$. From the character of the matrix of each symmetry class, we obtain $\Gamma = 3 \quad 0 \quad 1$. The character table for C_{3v} is given in the Solution to Problem 7.5, and we find $\Gamma = A_1 + E$.

3. Under C_{3v}, the atomic orbitals for the A_1 and E irreducible representations are

$$A_1: \qquad s, p_z, d_{z^2}$$
$$E: \qquad p_x, p_y, d_{xz}, d_{yz}$$

Possible combinations leading to three N–H bonds are

Trigonal planar:	sp^2, p^2d, sd^2, d^3
Trigonal pyramidal:	p^3, pd^2
Unsymmetrical planar:	spd

For a first-row element, we can exclude contributions from d-orbitals, which leaves sp^2 and p^3. Since H–$\widehat{\text{N}}$–H is 107°, a linear combination such as $\psi = c_1\phi(sp^2) + c_2\phi(p^3)$, would prove a suitable trial wave function for ammonia.

4. See Fig. S8.1.

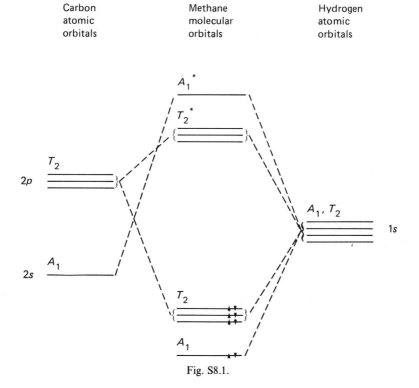

Fig. S8.1.

5. Using the Ni–F bond vectors as a basis, we generate

D_{4h}	I	$2C_4$	C_2	$2C_2'$	$2C_2''$	C_i	$2S_4$	σ_h	$2\sigma_v$	$2\sigma_d$
Γ	4	0	0	2	0	0	0	4	2	0

which reduces to $\Gamma = A_{1g} + B_{1g} + E_u$. The atomic orbitals thus available are s or d_z^2, $d_{x^2-y^2}$, p_x and p_y, so that possible hybrid orbitals are sp^2d or d^2p^2. Further calculations would be needed to distinguish between these two hybrids. (In fact, we know that sp^2d is preferred.)

6. Methane belongs to point group T_d, and has the irreducible representations $A_1 + T_2$. The wave functions are compounded from $C(sp^3)$ and $H(s)$ orbitals (see Fig. S8.1). Consider a transition $T_2 \rightarrow A_1^*$. $T_2 \cdot A_1 = T_2$. Consider the integral $\int \psi_1 x \psi_2 \, d\tau$, where ψ_1 belongs to A_1 and ψ_2 to T_2. Since x belongs to T_2, the product $\psi_1 x \psi_2$ is even, and the transition is permitted. If, instead, ψ_2 belongs to A_1 while ψ_1 remains as A_1, then $A_1 x A_1$ is even and $\psi_1 x \psi_2$ is odd, so that the transition $A_1 \rightarrow A_1^*$ is forbidden.

7. The z axis lies through N and normal to the plane of the three H atoms; x and y are in the plane of the H atoms, with x passing through one of them, and x, y and z making a right-handed set. Then, with the aid of Appendix 5, we can determine the matrices for the symmetry operations of C_{3v}:

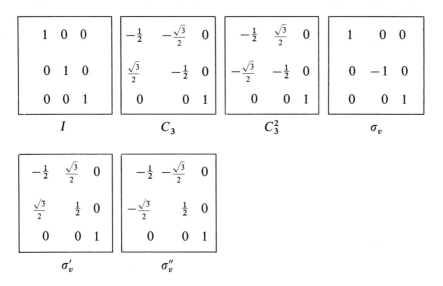

The traces of these matrices are 3, 0, 0, 1, 1 and 1 respectively. Hence, the representation for $I, 2C_3, 3\sigma_v$ is 3   0   1, which reduces in C_{3v} to $A_1 + E$. Hence, p_z belongs to A_1, and p_x, p_y to E. Thus, the three p-orbitals that are triply degenerate under spherical symmetry (the atom), are split into p_z (non-degenerate) and p_z, p_y (doubly degenerate) under the C_{3v} symmetry of the molecule—an example of ligand-field theory with p functions.

Bibliography

F. A. Cotton, *Chemical Applications of Group Theory* (1971) Wiley, New York.

L. H. Hall, *Group Theory and Symmetry in Chemistry* (1969) McGraw-Hill, New York.

N. H. Hartshorne and A. Stuart, *Crystals and the Polarization Microscope* (Arnold, 1970)

International Union of Crystallography, *International Tables for Crystallography*, Volume A (1983) Reidel, Dordrecht [*formerly International Tables for X-Ray Crystallography*, Volume I (1965) Kynoch Press, Birmingham].

M. F. C. Ladd and R. A. Palmer, *Structure Determination by X-Ray Crystallography* (1965, 2nd edn) Plenum Press, New York.

C. H. Macgillavry, *Symmetry Aspects of M. C. Escher's Periodic Drawings* (1965) Oosthoek, Utrecht.

F. C. Phillips, *An Introduction to Crystallography* (1971, 4th edn) Longman, Harlow.

A. E. H. Tutton, *Crystallography and Practical Crystal Measurement* (1911) Macmillan, London.

M. M. Woolfson, *An Introduction to X-Ray Crystallography* (1970) Cambridge University Press.

Index

Index